Geologic
Well
Log
Analysis

Sylvain J. Pirson
Professor of Petroleum Engineering
University of Texas
Austin

GEOLOGIC
WELL
LOG
ANALYSIS

Gulf Publishing Company
Houston, Texas

Dedicated to Meta, Dick and Jack

Geologic Well Log Analysis

Library of Congress Catalog Card Number 73-114682
ISBN 0-87201-901-2

Contents

Introduction   xi

1. SP and Eh Curves as Redoxomorphic Logs   1
Redox or Eh Potential; Significance of Redox Measurements in Exploration for Minerals; Basic Principles of Redox Log Interpretation; Future of Redox Logging in Oil, Gas Exploration; Mapping Problem.

2. Sedimentological Studies by Log Curve Shapes   37
Regressive, Transgressive Sedimentation Patterns; Offshore, Channel-Fill Sand Bars; Turbidites; Delta Sedimentation Sequence; Sedimentation Environment Mapping.

3. Exploration for Stratigraphic Traps   59
Recognition of Significant Geologic Features on Logs; Methodology; Characteristics of Common Stratigraphic Traps.

4. Continuous Dipmeter as a Structural Tool   84
Principles of Measurements; Continuous Dipmeters; Inaccuracies, Errors and Reduced Quality; Dipmeter Correlation and Computation; Using Dipmeter Surveys.

5. Continuous Dipmeter As a Sedimentation Tool   119
Fabric of Sedimentary Rocks; Petrophysical Properties; Determining Resistivity Anisotropy; Sedimentation Trend Studies from Conventional High Density Dipmeter Results; Azimuth Frequency Diagram; Modified Schmidt Diagram.

6. Paleo-facies Logging and Mapping   **153**
Lithology of Reefs, Facies Stratigraphic Traps; Porosity Log Combination; Computer Lithology Logging; Paleo-facies Mapping Techniques; Application.

7. Fracture Intensity Logging and Mapping   **180**
Fault Recognition; Log Characteristics of Turbidites, Slump Blocks; Subsurface Faulting; Fracture Intensity Mapping; Fracture Intensity Index; Fault Proximity Index; Fracture Density Index; Application.

8. Hydrogeology I: Hydrodynamics of Compaction   **207**
Geochemical, Radioactivity Anomalies; Hydrocarbon Leakage; Oil Entrapment by Hydrodynamics of Compaction; Radioactivity Mapping Techniques.

9. Hydrogeology II: Geostatic Equilibrium   **250**
Formation Pressures; Compaction, Fluid Relationships; Shale Compaction Effects on Electric Resistivity, Sonic Travel Time and Density; Well Logs in Estimating Formation Pressures; Application; Significance of Abnormal Pressures.

10. Hydrogeology III: Hydrodynamics of Infiltration   **284**
Hydrodynamic Oil Entrapment; Water Driving Forces; Well Logs as Hydrodynamic Potential Measuring Devices; Delineating Hydrodynamic Traps; Finding Hydrocarbon-Water Contacts; Hydrodynamic Flushing; Determining Fresh Water Flushing from Logs; Hydro-osmotic Studies in Thin Shaly Sands; Mapping Problems.

Appendix 1   **329**
FORTRAN II Language Program for Calculating the Dip Magnitude and Azimuth and the Degree and Orientation of the Resistivity Anisotropy.

Appendix 2   **339**
Lithology Computer Program: Three Minerals Plus Shale Contamination and Porosity.

Appendix 3   **357**
Computer Program for Calculating Structure Curvature.

Acknowledgments

The author wishes to thank various organizations for permitting reproduction of numerous illustrations from the original publications and for allowing adaptations of some of their published articles for this book. Specifically, the author wishes to mention with gratitude the Schlumberger organizations; the Dresser-Atlas Company; the American Association of Petroleum Geologists; the Society of Petroleum Engineers of the AIME; and the Society of Professional Well Log Analysts. These trade magazines have also contributed graciously to the illustrations and article adaptations: *World Oil* and the *Oil and Gas Journal.*

While preparing this book, the author was fortunate to present its contents as Continuing Education short courses to employees of the petroleum and of the mining industries under the auspices of the Extension Division of the University of Texas at Austin. In the course of these informal sessions, valuable exchanges of views took place with class members too numerous to mention here, and their anonymous contributions are hereby acknowledged.

Introduction

Heretofore, well logs have been used mostly for correlation, structural mapping and quantitative evaluation of hydrocarbon-bearing formations. The present series of discussion topics proposes to use well logs to their fullest extent for geological studies (sedimentation, fluid migration, tectonic deformation, etc.) by deriving from logs all sorts of mapping parameters and indices of significance from the point of view of mineral accumulations—primarily in sediments.

A basic knowledge and understanding of the physical measurements made in well bores is presumed of the reader and of students in addition to a good foundation in the principles of geology and—more particularly—of sedimentation.

In the first chapter, the SP curve is discussed as a redoxomorphic log, which indicates the various levels of oxidation or of reduction of sediments. The SP curve is viewed as a particular case of the new Redox potential log, which responds more directly to the reduction-oxidation states of the various formations. An application is shown in which the SP curve is used for mapping the extent of a potential hydrocarbon source bed and of its degree of devolatilization as a measure of the magnitude of the hydrocarbon volumes generated therefrom and of their location with respect to probable stratigraphic traps.

Chapter 2 deals with SP and resistivity curve shapes as indicators of the regressive and transgressive nature of sand deposition processes. Specific curve shapes for turbidites, delta sequences and sand bars are also discussed. A mapping problem is worked out by using a regression index, which permits to project laterally to a paleo-shoreline as the up-dip pinch-out of a regressive sand, i.e. to the probable position of a stratigraphic trap for hydrocarbons. Uranium ore accumulations are often associated with the down-dip pinch-out of regressive sands, and the same approach may be used to project laterally to the positions where the infiltrating uranium-bearing solutions may have precipitated commercial ore bodies.

Chapter 3 reviews using logs in prospecting for stratigraphic traps of all sorts (offshore sand bars, channel fill, valley fill, etc.) by selective isopach mapping of sedimentary genetic intervals below and between marker beds that are identified as geologic time lines. An example of application to the discovery of a probable new oil field on the trend of a known field (Black Jack Field, Denver-Julesburg Basin, Colorado) is worked out by three selective isopach maps.

Chapters 4 and 5 deal with the continuous dipmeter as a tool for structural, sedimentational and deformational studies. When dip angles and directions of dips are computed at a high density level (of the order of 20 computations per 100 ft.), patterns of dips and of their directions may be recognized which have meaning from a structural standpoint (anticline, faulting, drag, roll-over, etc.), from a sedimentation standpoint (sand bar, channel fill, draping, etc.) and relative to the positions of a well with respect to channel or sand bar geometry. This is of special usefulness in determining the direction in which a step-out well should be drilled in order to find as good or a better sand development as the one being studied. Means for determining from the dipmeter's correlation curves preferred directions of grain orientation as a primary sedimentation petrofabric or as a secondary tectonic petrofabric (because of lateral compression or of basement wrench faulting) are also presented with computer programs that permit the rapid solution of the equations.

Chapter 6 deals with lithologic logging making use of three porosity curves of high vertical resolution (density, sonic and neutron curves) together with the gamma ray curve for determining shale contamination. A computer program is provided for solving this problem (five unknowns at each level: three minerals, shale and porosity). Knowing the prevalent minerals is necessary in advance in order to use the adequate petrophysical parameters in the lithologic solution. An example is worked out for the San Andres reservoir, which is generally made up of dolomite, anhydrite and gypsum. The uses of this approach in evaluating sulfur and potash deposits are pointed out. Lithologic logging is of interest mostly in mapping lateral facies changes in carbonate-evaporite shelf deposits that lead to stratigraphic oil entrapment.

Chapter 7 is concerned with methods of mapping fracture development in brittle reservoir rocks such as chalk, limestone and dolomite. A fracture intensity index may generally be computed in such rocks when an adequate suite of electric logs is available. A problem is worked out for the Austin chalk in Central Texas. For this rock a correlation has been made between the fracture intensity index and the proximity to the major fault, Luling-Mexia. Similar correlations may no doubt be made in other geologic provinces to study fractures that major faults induced. Another fracture intensity mapping parameter may be derived from structural maps by computing the curvature of the deformed surfaces of stratification. An example of this technique is presented with a computer program.

The remaining Chapters 8, 9 and 10 deal with problems of *hydrodynamics*, which may be approached by well logs. One distinguishes two types of hydrodynamics: that of *compaction*, which starts the moment sediments are laid out; and that of *infiltration*, which takes place when the sediments are exposed at the surface of the earth and subjected to erosion.

Chapter 8 deals with using natural radioactive elements as tracers of fluid movement over geologic time. More specifically, this is done by mapping radioactivity anomalies within extensive and correlatable marker beds (shales, silts, etc.), as they reflect the variations in the cumulative vertical flux of formation waters expressed out by compaction under the increasing weight of the overlying sediments. Such anomalies have been shown to be related to the *primary* migration and pooling of hydrocarbons in favorable traps. It is possible to predict by subsurface radioactivity mapping why presumably good seismic structures are found dry because they failed to be in an area of convergence of the paleo-water flux. Examples of primary hydrocarbon pooling in structural and in stratigraphic traps are illustrated.

Chapter 9 deals with detecting overpressured formations (abnormally high reservoir pressures above the usual hydrostatic pressure gradient) by observing decreasing gradients with depth in shale density and in shale resistivity or by observing an increasing sonic shale travel time with depth. This problem is of increasing importance in deep oil and gas exploration—particularly in planning the drilling operations.

Chapter 10 deals with hydrodynamics of infiltration and with entrapment of hydrocarbons by water motion under pooled hydrocarbons. It is well known that such a phenomenon initiates tilted gas-water or oil-water contacts and that hydrocarbons may thus be swept out from their place of primary entrapment and migrate to secondary entrapment loci by migration en masse. Hydrodynamic entrapment is reviewed, and it is pointed out how determining the levels of 100% water saturation may contribute greatly to the confirmation of flow potential mapping. The oil-water contact is defined as the highest level at which Ro (resistivity at 100% formation water saturation) may be observed directly on logs. However, this is not always possible; and often it is necessary to project downward to such levels. This is possible even when wells do not reach the gas-water contact or water-oil contact provided the overlying capillary transition zone is partially penetrated. Examples of application illustrate the technique.

Chapter 10 deals also with connate waters' diluting by infiltration of meteoric waters from the outcrops. Dilution and break-through patterns around oil fields may be mapped by computing Rwe from the SP curves. Oil entrapment may also be explained by hydro-osmosis or by the creation of a shale pressure barrier according to Donnan's equilibrium. An example of oil pool mapping in the J sand, Denver-Julesburg Basin, is worked out by both methods.

**Geologic
Well
Log
Analysis**

SP and Eh Curves
as Redoxomorphic Logs

The SP curve was the first lithologic log that became available with the advent of electric logging. Yet, its interpretation and significance still are not understood fully. The SP curve was recognized only recently as a sedimentation curve; this significance stems from the SP curve's being just a subdued redox potential curve.

Redox well logging measures continuously the reduction-oxidation (redox or Eh) potential of the adjacent formations in a well bore filled with water-base mud (fresh or salty). This simple electronic potential measurement between two electrodes at the same level (an inert one, platinum; and a reference one, lead) indicates the proportions present in each geologic formation of oxidized and reduced forms of various minerals and ions in the formation waters. In sediments this potential depends on the deposition environment in which the sediments were laid out (deep to shallow neritic on the continental shelf, littoral and sublittoral along the shorelines of deposition); whether the winnowing action of the waves was a low or high energy environment; and whether the environment was modified by meteoric water, oil, gas, hydrogen sulfide, etc. when migrating and accumulating.

Evidence shows that the redox potential is an important component of the SP curve electric logs recorded.

Introduction

Economic accumulations of minerals in the earth are associated with definite and distinct geochemical environmental conditions. The numerous geophysicochemical parameters defining an environment may not be measured significantly all at once in a medium, albeit a geological medium. Those environmental parameters pertain to the rock-pore fluids in the aqueous phase. They only pertain to certain elements in solutions that are significant in controlling *insitu* chemical reactions and, more particularly, in controlling the equilibrium of the reactions.

The activities (concentrations) in various significant ions within the pore fluids of rocks define a geochemical environment. A Nernst potential representing the electron and proton transfer tendency within pore fluids (and at fluid-solid interfaces within pore spaces) expresses the determining factors of the reactions in the earth that have reached equilibrium under a prevailing environment. Such an environment may be relatively constant and uniform over certain geologic periods and may persist to the present. Our chief interest lies in geochemical reactions that occurred under relatively low pressures and at near surface temperature during rock genesis and later within sedimentary basins. Thus, we are concerned mainly with sedimentary geochemical reactions and equilibria.

The Redox Or Eh Potential

Oxidation-reduction reactions are essential in the development and decay of living organisms, as well as in solid mineral systems. Life processes are but continuous oxidation-reduction reactions.

Oxidation-Reduction Potential (ORP, redox potential or Eh) measures the tendency for chemical species in ionic solution to change their oxidation or reduction state. It also quantitatively measures the oxidation energy or the electron-escaping tendency of a reversible oxidation-reduction system.

Chemical reactions involving the transfer of electrons and protons (i.e., of electric charges) depend on the pH (proton activity measure) or the Eh (electron activity measure); or on pH and Eh of the systems in which the reactions occur. The following general expression governs such reactions at equilibrium and at standard conditions:

$$Eh = E_o - 60 \times \frac{a}{n} \times pH \qquad (1\text{-}1)$$

in which "a" is the protons (H^+) and "n" is the electrons (e^-) in the reaction. Equilibrium conditions are thus represented by a straight line on a Cartesian plot of Eh versus pH, the slope of which is $60a/n$. The above equilibrium reactions govern three chemical reactions:

1. Where protons are involved, as in this oxidation reaction:

$$Fe \text{ (solid)} + 2H^+ = Fe^{++} + H_2 \text{ (gas)} \qquad (1\text{-}2)$$

   (This *reaction* is pH *dependent* only. It is exemplified by acids, such as H_2SO_4, H_2S, H_2CO_3, HCl and H_3PO_4, reacting on metals.)

2. Where electrons are involved:

$$Fe^{++} = Fe^{+++} + e^- \qquad (1\text{-}3)$$

   (This is Eh dependent only. It is exemplified by the ionization of metals and oxidation from a low to high-valence metal ion.)

3. Where protons and electrons are involved:

$$Fe\,SO_4 + 2H_2O = SO_4^{--} + Fe\,O.OH + 3H^+ + e^- \qquad (1\text{-}4)$$

   (This reaction is Eh and pH dependent. Most reactions in the natural environment are like this.)

   In the last analysis, reduction always equals a net gain of electrons by the substance being reduced.

   Hence, a redox reaction involves transferring electrons, or the tendency to transfer electric charges. This tendency is measured electrically by a simple potentiometric circuit or high resistance millivoltmeter connected between a reference electrode (that supplies the driving voltage for transferring electrons) and an inert electrode, P_t or A_u, (that accepts the electrons). It measures the electric voltage they create.

The reference electrode is always construed to be the *hydrogen electrode*. Thus, Eh represents the redox potential. In practice, however, a hydrogen electrode is unwieldy and impractical. For laboratory and field surface work, a calomel electrode is used as the reference electrode at standard conditions of 25°C and one atmosphere, Ecalomel = – 245 mV.

Pressure compensated calomel electrodes may be used in wells; but, metallic electrodes are preferable if they are stable, free of poisoning and oxidation state changes under logging conditions (i.e., for the muds involved). Lead electrodes are satisfactory, whereas iron electrodes are not.

Knowing the behavior of a metallic reference electrode in a particular mud for varying temperatures and pressures with respect to a calomel or hydrogen electrode, it is possible to draw a zero line on a redox log. This is the first time that an SP-type log can be read from a significant zero line. This line will be established at the well site under logging conditions by immersing the logging electrodes in a solution of known redox potential, such as the ZoBell solution, (22) which has an Eh = 0.428 volt at 25° C.

The following shows the measured redox potential of a solution containing various species of ions in various oxidation and reduction states:

$$Eh = E_o + \frac{RT}{nF} \ln \frac{[Ox]}{[Red]} \tag{1-5}$$

where Eh is the redox potential exhibited by a pair of electrodes, closely spaced and dipped in the solution containing the oxidized [Ox] and the reduced [Red] ionized species of various elements and compounds. It is referred to the standard hydrogen electrode, which is set arbitrarily at potential zero.

E_o is a standard potential required to normalize the Eh values to the hydrogen electrode. Remaining symbols in the equation are identified accordingly:

R = gas constant : 8.315 joules
T = absolute temperature, °Kelvin
n = number of electrons involved in the reaction
F = 96,490 Coulombs

A solution's redox potential depends only on a ratio (Ox/Red), and adding ionized neutral salts such as Na Cl does not affect it. Hence, small quantities of formation waters mixed with mud in the well, even a salty mud, will give the same Eh potential as if undiluted or uncontaminated. This is the first time that an SP-like curve may be measured in salt muds.

The mud and formation water do not need the same composition to measure the formations' redox potential. However, the same ratio (Ox/Red) must be in the mud and formation at all levels. This is attained within a reasonable time after mud circulation is stopped because the SP currents in the well and formations rapidly transport the ionic charges necessary for this equilibrium.

When an inert electrode (platinum or gold) is immersed in a reversible oxidation-reduction system, a potential difference is set up at this electrode. A suitable potentiometer connected to a reference electrode inserted in the system readily measures this potential. The inert electrode, when immersed in the system, does not participate in any reaction but may act as a catalyst. It merely conducts electrons to and from the system. The inert electrode may be considered a reservoir of electrons of variable concentration—the potential of which varies with the electron concentration. Since the electron-escaping tendencies are different in the inert electrode and the redox system, an electric potential difference is set up at the Eh electrode.

Eh measures the *intensity level* and not the *capacity* of the electron-escaping tendency in the system. In this respect, Eh resembles temperature and pH, which are also intensity parameters. Respectively, they give no information about a system's heat or buffering capacity. Hence, the Eh of a system, which depends on the oxidant - reductant ratio, is independent of their absolute amounts and of the ease with which these amounts may be varied. The system's *poising capacity* measures the resistance to change in the reactants that are involved. For example, the Eh of an oxidized system may be the same whether it contains 1 or 10 percent oxidant. The resistance to change (i.e., the poising capacity), however, is ten times greater in the latter. Natural environments may have well-defined Eh potentials but be poorly poised—in which cases, it is difficult to obtain stable redox measurements. In the last analysis, all oxidative reactions remove electrons from the substance being oxidized. Since a system must remain electrically neutral, a reduction—the gaining of electrons—accompanies every oxidation. Hence, all oxidation-reduction reactions essentially consist of an equilibrium exchange of electrons between oxidants and reductants within the same system or environment. Oxidation cannot proceed without a corresponding oxidant to take up the liberated electrons. Oxidizing agents can take up electrons, and reducing agents can lose electrons. To measure the intensity level of a system's reducing or oxidizing ability, the electron-escaping tendency or the electron migration of electric charges must be measured. Thus, measuring electric potential differences does the quantitative study of the redox processes.

An environment's electron-escaping tendency is characterized by the Eh potential of an inert electrode and by the pH proton-escaping tendency, which is also an important environmental characteristic.

Significance of Redox Measurements in Exploration
For Oil and Gas and for Other Minerals

Redox measurements are signigicant in oil, gas and other mineral exploration because they measure the physicochemical factors at work during sedimentary diagenesis. Internal changes in the sediments occurred from the time they were laid out until the present. The following are the diagenetic periods according to Dapples[5]:

1. In early or redoxomorphic diagenesis, the action of oxygen in sea water is particularly active and involves mostly iron compounds. Thus, during this period, the color of sediments changes.

2. During late or locomorphic and phylomorphic diagenesis, sediment cementation occurs because of a shift in electrochemical equilibrium. Mineral replacement is also an active process.

3. In present or epigenetic diagenesis, shallow mineralization occurs because of leaching of elements from a protore with transportation and reprecipitation in favorable physicochemical environments. This process is particularly important in the genesis of sedimentary deposits of uranium, vanadium, manganese, iron, copper, etc.

Certain pertinent quotes illustrate these diagenetic processes:

According to Larsen and Chilingar,[15] "diagenetic transformations may be characterized collectively as the reactions of the sediment to its physicochemical environment, the most important factors of which are pH and Eh combined, ionic concentration, temperature and pressure of the environment."

According to Amstutz et al.,[1] "the chemical characteristics established during diagenesis are preserved for a long time if the sedimentary deposits have not been subjected to weathering, recent or ancient." He added, "the diagenetic environment consists of two phases, a solid or detrital phase the components of which are oxidized to various degrees (the most important being iron) and a liquid phase contained in the pore space which originally is practically always sea water. The latter's main characteristics are its salts, its organic matter in solution or suspension, living organisms and dissolved oxygen. As a result of consuming oxygen, the environment becomes progressively more reducing, which is usually associated with a notable change in pH through temporary acidification."

Figure 1-1[4] diagrams physicochemical zones related to carbonate sediments in which the Eh-pH relations are indicated.

Figure 1-2[7] is an idealized profile through a continental margin indicating the three phases of diagenesis.

Figure 1-3[8] is a generalized Eh-pH diagram illustrating the limits of natural environments.

Figure 1-4[13] is a fence diagram that separates the principal environments of sedimentation and diagenesis together with the probable minerals expected in each.

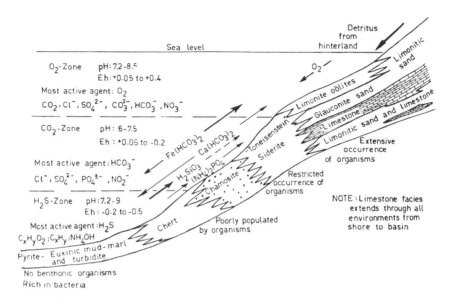

Figure 1-1. Diagrammatic presentation of physicochemical zones related to diagenesis of carbonate sediments. (Courtesy of G. V. Chilingar and Elsevier Publishing Company.)

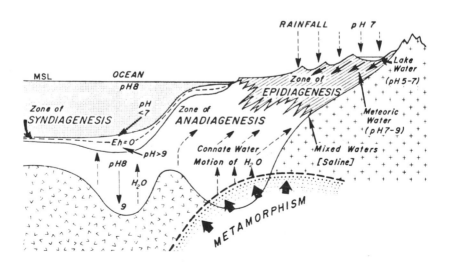

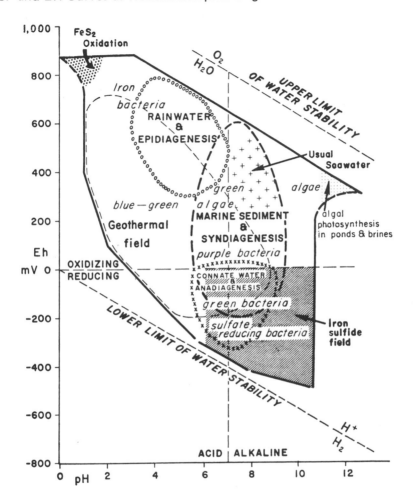

Figure 1-3. Catenary diagram illustrating limits of natural environment in terms of pH and Eh—especially, the sites of syn-, ana- and epidia-genesis. (Based on works by L. G. M. Baas Becking and R. M. Garrels.) (Courtesy of R. W. Fairbridge and Elsevier Publishing Company.)

Figure 1-2 (bottom left). Idealized profile through a continental margin showing the sites of contemporary marine sedimentation and the three phases of diagenesis: (1) diffusion potential during syndiagenesis; (2) upward liquid motion in anadiagenesis; (3) downward motion in epidiagenesis. (Courtesy of R. W. Fairbridge and Elsevier Publishing Company.)

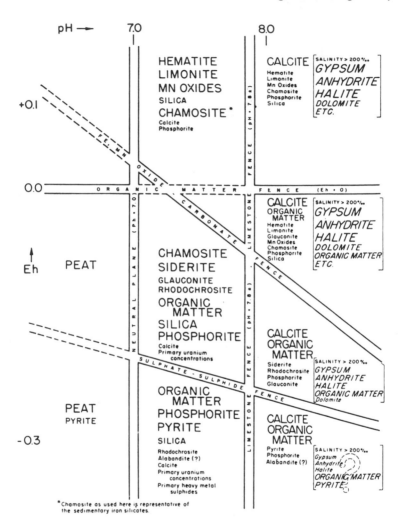

Figure 1-4. "Fence diagram" illustrating principal environments of sedimentation and diagenesis according to Eh and pH. (Courtesy of W. C. Krumbein, R. M. Garrels and *The Journal of Geology*.)

Figure 1-5[6] gives the limits of realms of clay minerals and micas as environmental indicators as a function of Eh and pH.

According to Hunt,[11] "most oil forms in marine sediments because these are more reducing than continental beds. . . . Hydrocarbons old and new will be preserved if they are in a reducing environment. . . . Deeply buried organic matter will form more petroleum than organic matter remaining near the surface." These observations show the redox log's potentials for evaluating source beds and mapping their great extension.

Basic Principles of Redox Log Interpretation

The fundamental principle of redox log interpretation is that the redox potential measured by the log responds to the ratio of oxidant to reductant substances within the formations traversed

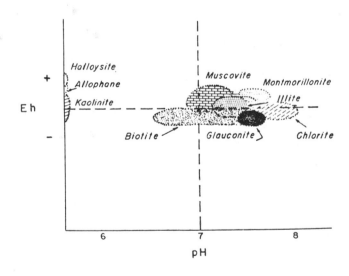

Figure 1-5. Generalized stability realms of clay minerals and micas based on paragenetic sequence of associated minerals and miscellaneous environmental indicators. The intent is to show Eh and pH conditions under which a clay mineral or mica assemblage tend to form preferentially in sandstones. (Courtesy of E. C. Dapples and Elsevier Publishing Company.)

by a bore hole, assuming that the mud in the hole has come in redox equilibrium with the formations before running the redox probe into the hole. At the various discrete sedimentary levels, the recorded Eh potential should then be a measure of the oxidation degree or of sediment reduction, assuming that no later sediment modification by diagenesis or by fluid motion has occurred. When the latter happens, effects may be evaluated by comparison with unmodified environments. The reason for the modifications may then be evaluated and found to result from oil and gas accumulations, fresh water dilution, etcetera.

Sediments laid down from a shoreline to the edge of the continental shelves vary in oxidation degree and, therefore, in the Eh values that might be recorded in wells drilled through them. Table 1-1 shows the redoxomorphic series that might be encountered.

Table 1-1
Sedimentary Redoxomorphic Series

Sedimentation Environment	Relative Oxidation Level and Water Depth	Approximate Eh Range Millivolts
Littoral	Lagoonal	+ 500 to + 600
	Beach sand and sand bar	+ 600 to + 650
Sublittoral	Offshore sand bar (0-50')	+ 500 to + 550
Neritic	Shallow - (50-200')	+ 400 to + 500
	Medium - (200-400')	+ 400 to + 450
	Deep - (400-600')	+ 375 to + 400

The superposition sequence of these sediments is informative concerning the nature of the offshore sedimentation—as a result of a regressive or transgressive shoreline. Some hypothetical examples will permit a better understanding of the interpretation techniques.

Sedimentation Sequences Associated with a Regressive Shoreline

Figure 1-6 schematically represents the sedimentation sequence associated with a regressive shoreline as the sea water level is successively lowered from 1 to 2 to 3 relative to the land surface, gradually exposing more continental shelf to disintegration by wave action. At each of the three successive stages sediments which are coarse and free of shale near the shoreline are laid out. They are well oxidized, and the wave energy that winnowed the sand grains by its to-and-fro action and tide motion remove the clay particles. The apron of the nearshore sand gradually tapers seaward into shallow neritic sediments that are mostly shaly sands and silts laid out in a low energy environment which is only slightly oxidized.

Further down the continental shelf, the shallow neritic sediments grade laterally into deep neritic shales in a reduced environment of very low wave energy. As the sea level drops from 1 to 2 to 3, similar sediments are successively laid out with an off-lap in a vertical position down-dip along the continental shelf slope. After the above sedimentary sequences are further buried and compacted, formations waters may be expelled. The redox potential, however, is not necessarily modified and may be ascertained by the redox logs from wells such as I, II and III.

The hypothetical Eh curves that would be recorded are shown for each well at the corresponding levels of the sedimentary sequence of beds 1, 2, 3 according to their respective positions in the redoxomorphic series. All three Eh curves exhibit the regressive feature generally observed on SP curves—a displacement upward toward more negative values within the sedimentation sequence. This Eh displacement may be 50 to 150 mv negative in clean sands laid out in a high energy oxidizing environment and may be down to 5 to 15 mv negative in the deep neritic shales. In

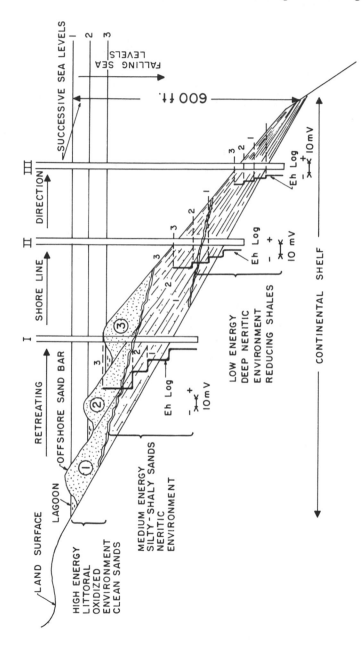

Figure 1-6. Schematic representation of sedimentation environments associated with a regressive shoreline. (Courtesy of *The Oil and Gas Journal*.)

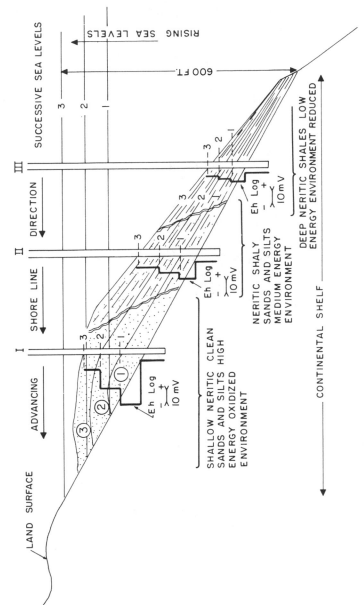

Figure 1-7. Schematic representation of sedimentation environments associated with a transgressive shoreline. (Courtesy of *The Oil and Gas Journal*.).

the latter case, only a shale baseline shift toward the negative is observed, as shales are successively laid out in waters slightly richer in oxygen.

Such shifts and gradients on the actual redox log of Figure 1-8 are quite evident and are noted.

Redox well log readings are not necessarily the ones that prevailed at the time of sedimentation because the sediments had been subjected to intense bacterial action soon after they were laid out and to diagenetic changes over geologic time. Significant relative displacement relationships, nevertheless, might have been preserved.

On the actual redox log (Figure 1-8), if the SP curve were amplified and if the redox curve were shifted upward 5 to 6 feet (a displacement inadvertently introduced while logging), the two curves could be matched almost exactly—at least so far as the main SP deflections are concerned. This is as it should be if the Eh curve is the "open battery potential" that generated the SP current, since the ohmic potential (ir) drop along the bore hole is recorded as the SP curve. However, this adjustment will not modify conclusions that the SP curve should indicate residual oil, wateroil contact and various oxidation levels at the neritic zones of sedimentation in the shale sections. It is possible to adjust and determine such shifts from the SP curve. Several parameters *Projective Well Log Interpretation* uses are so derived.[18]

Sedimentation Sequences
Associated with a Transgressive Shoreline

Figure 1-7 schematically represents the sedimentation sequence associated with a transgressive shoreline as the sea level rises successively from 1 to 2 to 3, relative to the land surface as the shoreline advances gradually over the continental shelf and deposits a continuously onlapping coarse sand. At each of the 1, 2, 3 stages, sediments which are coarse, relatively free of shale and well-aerated (oxidized) near the shoreline are laid out. The wave energy that winnowed the sand grains removes the clay particles. In essence, the vertical and lateral sequence of sediments resembles that previously discussed for a regressive shoreline. The major difference is that the successively younger beds are laid out in waters that are gradually becoming deeper. Accordingly, at a given point, such as, in a well drilled much later—the oxidation level the Eh log measures will gradually decrease upward (i.e., the redox curve will swing to the right toward positive deflections of the conventional SP curve).

The hypothetical Eh curves that would be recorded from transgressive sediments are shown for wells I, II, and III with appropriate deflections at corresponding levels of the sedimentary sequence of transgressive beds 1, 2, and 3 according to their respective position in the redoxomorphic series. The expected Eh displacements and shale base line shifts are represented in magnitude and sign at the three wells. The actual redox log of Figure 1-8 again shows some well-marked transgressive features.

First U.S. Redox Well Log

Figure 1-8 reproduces the first redox log run in the U.S. with a probe supplied by C.E.D. (Carotaggio Elettrico Differenziale) of Bologna, Italy. Four metallic electrodes on the probe (2Pb, Pt, Au) allowed continuous passage of the mud in the hole over the electrodes to provide a continuous cleaning action. For reasons beyond the logging company's control , logging had to run at 100 feet per minute. Even at this unusually high speed, duplicate runs of the redox log matched each other with unexpected accuracy. Accordingly, Figure 1-8 shows only one run. We note again that the redox log was displaced inadvertently 5 feet downward with respect to the conventional SP curve. Tracts II and III show the conventional curves of the induction log survey.

The SP curve (Tract I—solid line) comes from the lead electrode on the redox probe, whereas the redox curve (Tract I— dashed line) measures and records the potential between the same lead electrode and a platinum electrode at the same level. Hence, no ohmic effect (or potential drop due to the so-called "SP current") can be involved in the redox curve. The redox curve deflections, recorded on the same scale as the SP curve, are much larger than those of the SP at corresponding levels.

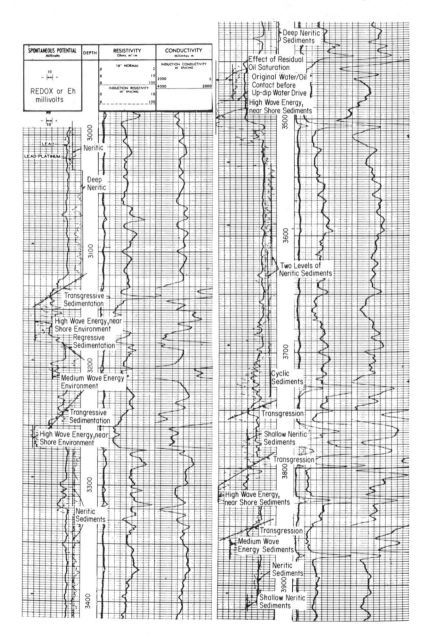

Figure 1-8. First redox well log run in the United States. (Courtesy of Dresser-Atlas Company.)

The hypothesis states that the redox potential is related to the true open circuit potential (i.e., without supplying any current) in the well bore. The author presented this hypothesis in a letter (April 24, 1943) to Dr. Claude E. Zobell of the Scripps Institution of Oceanography in La Jolla, California. The following paragraph appeared in this letter: "I am still wondering if what the electrical log (SP curve) is measuring is not the difference in Redox potential between formations." This statement also referred to the author's discussion with Dr. ZoBell at the AAPG (American Association of Petroleum Geologists) meeting in Fort Worth, Texas, a few weeks earlier. While the author, his students and associates continued working along these lines, no conclusive evidence was ever produced until the redox log of Figure 1-8 became available. This log indicates that the main source of potential in a well is the redox potential and that the SP curve is merely what remains when the "Eh cell" produces an SP current in the earth thereby lowering the magnitude of deflections recorded as the SP curve.

Redox Potential Log as an Open-Circuit SP Log

When mud circulation is stopped prior to well logging, fluids of different redox states are moved in and around the well bore. A schematic representation in Figure 1-9 shows what might be occurring. The following give the individual potentials:

$$Eh_m = E_{o_m} + \frac{RT}{nF} \ln \frac{Ox_m}{Red_m}$$

$$Eh_{sh} = E_{o_{sh}} + \frac{RT}{nF} \ln \frac{Ox_{sh}}{Red_{sh}}$$

$$Eh_o = E_{o_o} + \frac{RT}{nF} \ln \frac{Ox_o}{Red_o}$$

$$Eh_{sw} = E_{o_{sw}} + \frac{RT}{nF} \ln \frac{Ox_{sw}}{Red_{sw}}$$

$$(1-6)$$

When two redox systems are placed in contact through the intermediary of a semipermeable membrane, electric charges trans-

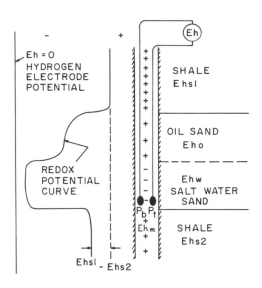

Figure 1-9. Distribution of net electric charges in a well bore drilled with fresh or salt mud as a result of redox potential log.

fer, resulting from placing the various redox systems in contact and—more particularly—when the mud system is not well poised (i.e., when the system may change appreciably in net electric charge per unit volume). Since shales are well poised (i.e., since they have a large capacity to exchange electrons, protons and other ions), they will modify the electric charge of the mud facing them by oxidizing it—by absorbing mostly chloride ions from the mud. The mud thus acquires positive charges in front of shales. To a lesser extent, the same result occurs in front of the oil zone where the mud acquires an excess of positive charges. In front of the water zone, the mud recieves extra negative charges because the water zone in the main acts as an oxidized zone. When equilibrium sets, charges and potential are distributed as in Figure 1-9.

Considering the Eh potential the oil zone creates, the oil effect is well marked and the wateroil contact level can be observed without question. The actual log in Figure 1-8 shows this effect at 3,484 feet on the redox curve, whereas it it not visible on the SP curve. The contact may be observed, however, from the induction

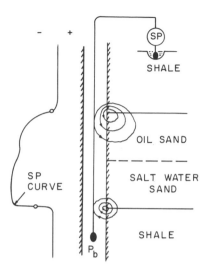

Figure 1-10. Distribution of SP currents in a well bore and resulting SP curve that smooth the significant stratigraphic redox potential levels.

log as a small difference in level between resistivity and conductivity curves.

The SP curve, which results from measuring the ohmic potential drop along the well bore because of formation electric currents the charge distribution generated (Figure 1-10), is only a subdued measurement of the redox potential distribution as it exists.

The redox log also shows various discrete levels within shale sections that indicate the deposition environment. Such levels are highly subdued or unobservable on the SP curve.

Transference of charges and equilibrium (Figure 1-9) will occur if the mud in the well is not well poised, i.e., if it has a low resistance to an equilibrium change. This applies to fresh water drilling muds that contain little or no organic substances. Presently, the tendency in well drilling is to drill with fresh water to great depths or until weighing material or stabilizing and mud-thinning chemicals must be added. Under such conditions, redox logging will give excellent and significant results. It will occur when sea water is used as drilling fluid on offshore wells.

Adding organic chemicals such as quebracho, starch, carboxylmethylcellulose and oil in oil emulsion muds reduces the mud in character and renders it well-poised on the reducing side. It may never come in physicochemical equilibrium with the formations drilled through, and a significant redox log may not be obtained unless the mud is oxidized or circulated out and replaced by fresh mud or salt water.

Under certain conditions, the conventional SP curve acts as a redox log. This particularly applies when the reservoir rocks are shaly sands, where the SP deflections are highly compressed in the oil zone when compared to the water zone deflections. Two such logs, both from the Holly Ridge Field, Louisiana, are in Figure 1-11. The water-oil contact at 8,380 feet in Log 1 is overlain by a well-marked SP gradient of 40 millivolts over 24 feet. In log 2, the water-oil contact at 8,420 feet is overlain by a similar SP gradient of 16 millivolts over 20 feet.

Both logs show that the mud filtrate invasion is very shallow, and—therefore—the mud in the wells responds to the redox effect of the oil in formations around the well bore. Kerver and Prokop[12] have studied this effect, observing similar reduction in SP values with formation water saturation reduction from 60 to 10 percent in shaly sands and in clean quartz cores. They base their explanation on the double layer effect rather than redox potential.

Water-oil and water-gas contacts are also difficult to determine in many other instances in sandstone type reservoirs, especially when the formation porosities are low and when long capillary transition zones are present. Contacts would become obvious with redox logs run in fresh and salt muds.

The redox log in Figure 1-12 was from the Candela No. 16 Well, Foggia Field, Italy. Redox log in this well is the difference between a regular SP curve run and an iron electrode placed on the induction log sonde during the regular IES survey. After this survey, another SP-type log was run, using a gold electrode in the well plus the surface reference electrode used in the IES survey. A gold electrode acts as the inert or noble metal electrode required for redox measurements.

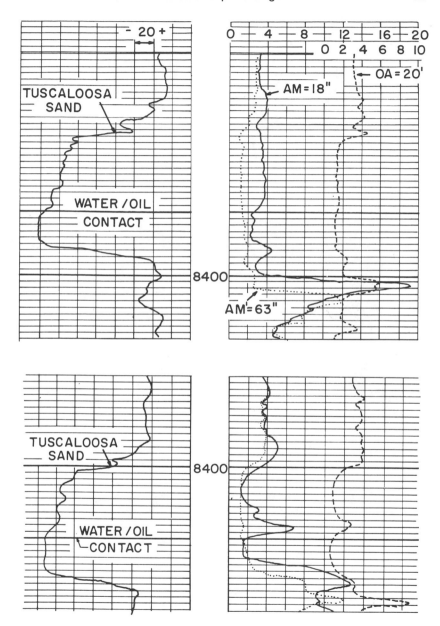

Figure 1-11. Examples of SP reduction caused by oil in a shaly sand—Basal Tuscaloosa, Holly Ridge Field, Louisiana.

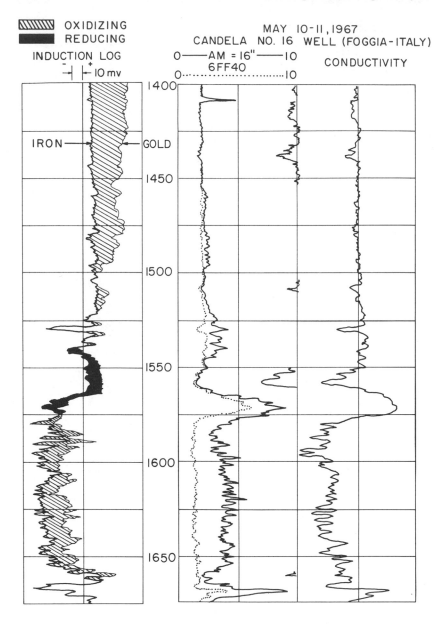

Figure 1-12. Differential SP obtained from two SP logs: one with an iron electrode; the other, a gold electrode. (Courtesy of Veneziani and *The Oil and Gas Journal.*)

The two records of the same sensitivity have been super-imposed on the SP side of the log in Figure 1-12. A well-defined water-bearing zone runs from 1,575 to 1,660 meters; the hydro-carbon-bearing zone, from 1,565 to 1,575 meters. In the waterzone, the differential reading between the iron and gold elec-trodes averages 10 to 20 millivolts positive, indicating that the water-bearing zone is oxidizing with respect to the oil zone. Where the differential is negative, it indicates a reducing environment. In addition, the shale section (1,540 to 1,565 meters) is a reducing environment, which may be considered a source bed for oil and gas contained in the reservoir.

Determining Fresh Salt Water Contact

Since fresh ground water comes from aerated meteoric water infiltration, a sharp Eh contrast between connate salt water and fresh water should exist. However, as a result of ground water percolating downward, some oxygen gradually disappears with infiltration distance. The contact may thus become a transition zone, or an Eh gradient, from an oxidizing environment in the fresh water zone to a less oxidizing (or even reducing) environment in the connate (salt) water zone.

A major problem of formation evaluation by well logs is finding oil and gas in the transition zone from fresh water to salt water, especially where the salinity change is not well marked and may not be evaluated from the SP curve. Numerous examples of these problems are in the Gulf Coast of South Texas where the Carrizo Sand (a notorious fresh water sand) overlies the Wilcox Sands, which are good oil and gas prospects, often in thin zones. There the SP curve is not appropriate for evaluating water resis-tivity with the accuracy required for quantitative formation evaluation. If properly run, the redox log would pick out hydro-carbons without difficulty in all such sands—thin or thick, shaly or clean, with shallow or deep invasion, containing fresh water or salt water.

Finding the exact water-oil and water-gas contact level is very important in reserve estimates, in unitization dealings where a participation plan must be worked out on an equitable basis, in zoning reservoirs for engineering and hydrodynamics studies; and for projecting fluid contacts laterally and for well completion (i.e., perforation and cementing).

An SP log on offshore wells is particularly difficult to obtain because of the moving and unstable reference electrode. In redox logging no such difficulty appears, since the two electrodes are at the same level in the well bore. Wallace[20] has shown examples of particularly troublesome SP curves.

Another application of this logging would be to trace the infiltration distance of polluted water into the ground water supply, as water polluted by organic matter exhibits a negative Eh, which may increase gradually with distance from infiltration points as the pollutants are absorbed or destroyed.

Determining the Crude to Be Expected
and the Proximity to Oil and Gas Accumulations

Vdovykin[19] has shown that a direct relationship exists between the oxidation-reduction state of formation waters and hydrocarbons contained in the rocks from which the waters were obtained. Figure 1-13 includes the data from Vdovykin, Germanov[10] and the writer wherein the physicochemical properties (Eh and pH) of formation waters are cross plotted. The Cis-Caucasian oil field waters have their representative points in a restricted cluster "A", that indicates alkaline-reducing conditions: Eh : + 10 to + 180 mv; pH: 6.8 to 8.3, whereas the formation waters from nonproducing areas lie above cluster A in a restricted area B, which is more alkaline (pH: 8.0 to 9.0) and more oxidizing (Eh : + 200 to + 260 mv). Some of these waters had been filtered; i.e., they were exposed to air. The waters represented by cluster "C" are from sea, estuaries and lakes. Oil samples and water containing H_2S in solution had negative Eh values, which are not on the graph. Luling and Lytton Springs waters show well-separated points on the Eh-pH cross plot; this technique might characterize various formation waters and their associated crude oils.

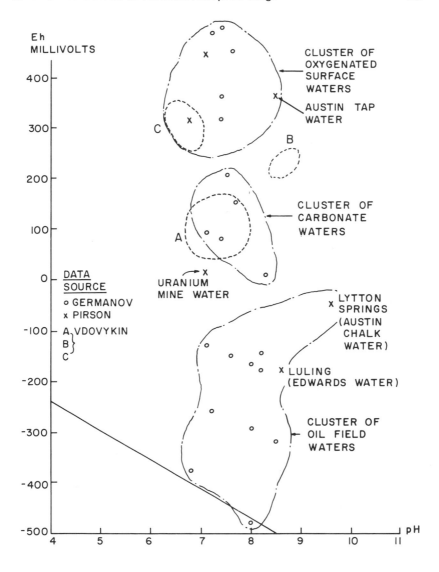

Figure 1-13. Formation waters characterized by their Eh-pH. Combining data from Vdovykin, Germanov and Pirson. (Courtesy of *The Oil and Gas Journal*.)

Weeks[21] reports that the environmental conditions of the source beds and their facies substantially influence the crude oils derived from them. Table 1-2 is based on Weeks' observations.

Table 1-2

Crude Oil Types	Preferred Environment and Facies
Asphalt base	Marine carbonates — neutral Eh
Asphalt base plus sulfur	Marine carbonates — low Eh
Paraffin base	Semimarine to brackish — deltaic and sublittoral sand and shales Positive Eh
Paraffin base with high wax content	Nonmarine — Continental sands and shales, Coals High positive Eh

It appears, therefore, that redox logging will permit evaluating the crude oil type to be expected in a given petroliferous basin and the proximity to oil accumulations when enough of these logs become available.

The Future of Redox Logging in Oil and Gas Exploration

Considering the oil and gas occurence statistics of commercial accumulations within structural and stratigraphic traps, overwhelming evidence shows that sedimentary environment is more important than trap geometry and trap occurence as a determining factor of actual entrapment. From a sedimentary environment standpoint, only two requirements that justify its petroliferous nature are necessary: (1) abundant organic life to confirm the expectation of rich organic source beds; (2) rapid subsidence of these sediments to prevent the complete decay of life remains before burial. Organic source beds are found in the stagnant areas of sedimentary basins, where a reduced environment is created in the absence of an oxygen supply. Other organically rich sediments may be found in *delta margins* as delta plain sediments and fans, salt marshes, flood plains—where rivers supply rich organic matter in suspension. Because of the sedimentary load, an overburden pressure is rapidly exerted on the edge of a basin by a prograding delta; subsidence of the region is rapid; and preserving organic matter as future source for oil and gas generation is assured.

On the *sandy shelf margins* of basins, favorable source bed environments are generated behind barrier beaches and islands, in lagoons and in bays. If they are rapidly sinking, the required source beds may again be preserved. Similar conditions are required on *carbonate shelf margins*, where organic life environment is generated in the backreef lagoons as well as in the reef proper and forereef detrital material. Again, for future oil and gas generation, a quick burial of the organic sediments allows oil and gas to migrate and accumulate in the patch reefs and stratigraphic traps of the back reef lagoonal sediments, in the reefcore proper and forereef detrital traps.

Another environment favorable to oil-gas generation and accumulation is in *turbidite sands* that rapid gravity suspension flows of shallow sedimentary loads released from gravitational instability on the steep slopes of continental shelves (generally in delta regions) produces. Such turbidites which are rich in nonoxidized organic matter, preserve the character of shallow water sedimentary environment and form deep water fans. Thus, their water characteristics are of deep ocean waters; their grain characteristics, of shallow waters.

The principal petroliferous environments previously discussed are all characterized by organic matter—therefore, by a definite low Eh value recognizable on redox logs. Only future experience based on many logs and correlation with known petroliferous environments will permit delineating the favorable environments and proximity thereto. The *rapid burial* requirement of the organic sediments may also be verified from redox logs. Geologists recognize this feature by the vertical rapidity of rock color changes above source beds. Rapid sedimentation is associated generally with an environment becoming more reduced at shallow depths as would occur when land subsides under a transgressive sequence of beds, forming strata that prograde toward land as sea level rises.

A positive *vertical redox gradient* may thus be considered a measure of the sedimentation rapidity by a transgressing sea or by the sinking of a basin's margin. Such an effect quite possibly is recorded on the Eh log of Figure 1-12 between 1,400 and 1,500 meters (deep). In this interval the potential recorded by the redox electrode (gold) departs gradually from the conventional shale base line (iron electrode), indicating that a more reduced environment is developing toward shallower depths, (i.e., the shales in this section gradually are being deposited in a transgressive sea). A transgressive sedimentary overburden is not as propitious as a regressive environment to preserving organic matter in underlying source beds. This may explain why the hydrocarbon-bearing section—1,565 and 1,577 meters—is not a commercial accumulation. Besides hydrocarbon gases, this interval produced considerable carbon dioxide. A reversal in the redox curve's vertical gradient would have indicated a regressive sedimentary overburden and environmental conditions over the source bed (1,540 to 1,565 meters), which would have been much more propitious to preserving organic matter and transforming it into a commercial hydrocarbon accumulation below.

This interpretation is valid if the Eh gradients or shifts are not produced by freshsalt water transition zones or contacts which could have been verified by a salinity mudlog while drilling the well.

Mapping Problem 1
Mapping Source Beds and Their Areal Degree of Depletion

Essentially, the devolatilization degree of source beds may be considered a measure of the hydrocarbon mass they might have generated over geologic time. Source beds are considered mainly to be shales; but, it is likely that lignite, bituminous coal and even organic limestones may have acted as source beds. Organic matter in such beds was responsible for their original Eh values on the reduced side. With devolatilization, the Eh gradually increased; the higher the Eh, the higher the degree of devolatilization (i.e., the more hydrocarbons that have been generated).

For these hydrocarbons to form commercial accumulations, they must first migrate and then become entrapped in geologic containers (reservoir rocks) from which they may not be dislodged.

The foremost hydrocarbon migration agent is formation water, expressed out by compression from compactible shales, that moves laterally toward land under the overburden pressure exerted by gradually compacting sediments. The compacting path of the paleohydrodynamics must therefore be ascertained and, in this path hydrocarbon traps (structural, stratigraphic, hydrodynamic or hydro-osmotic) that will accumulate and retain the pooled hydrocarbons must exist.

Redoxomorphic logs may help solve the areal distribution of source beds and their depletion, whereas other types of logs will help solve the problems of trap existence, geometry, porosity development, fluid migration and reservoir energy.

We will study now how to use the SP and Eh curves to map source beds and to ascertain their degree of devolatilization.

When electric logs are available over a large prospective area, and when sand developments most likely to act as reservoirs are lenticular and associated with a dominant regressive shoreline, the most probable source beds for such traps are the rapidly transgressing shales that cover the regressive sand series. The relative redox potential level presently in the overlying source beds may then be measured by referring to the potential level of more deeply buried shales believed to be nondevolatilized. An example of such a source bed mapping procedure is in Figure 1-14 for the shale overlying the J sand in the Denver-Julesburg basin (in the D-J interval). This shale is devolatilized over the Black Jack Field, Sections 9 and 10, which produces from the J sand; other regions

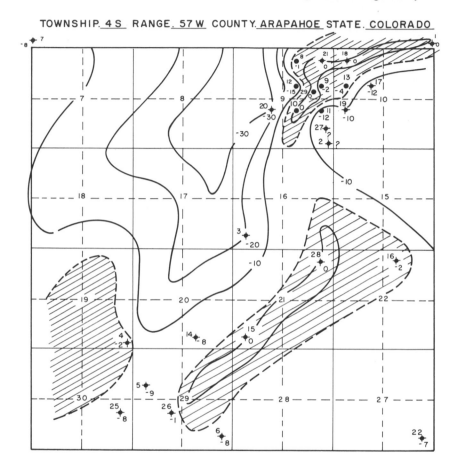

Figure 1-14. Map of the relative devolatilization of a source bed—i.e., the shale between the D and J sands, Denver-Julesburg Basin. Hachured areas have generated most of the hydrocarbons.

in Sections 21 and 19 and 30 indicate that hydrocarbon generation has also occurred.

This method, of course, is highly empirical presently. It depends on the deeper shale's constant properties; on the drilling muds not being organic (quebracho); and on the SP electrode's being the same throughout the various surveys and being stable under mud composition, temperature and pressure conditions.

Should Eh logs have been available under proper zero calibration and proper mud conditions, the redox state of the source beds could have been mapped in an absolute manner.

Conclusions

The environmental logging concepts of the redox log open these vast horizons for research and field applications, which are not in conventional logging methods:

1. Immediately, the redox log will be particularly useful in detecting hydrocarbons and hydrogen sulfide in reservoir rocks. It permits obtaining a significant SP-like curve in salt muds. Positive identification of water hydrocarbon contacts is a major improvement.

2. In *future* hydrocarbon and hydrogen sulfide explorations—when numerous redox logs become available—the sedimentary character, environmental conditions and fluid motion patterns, will be derived from redox logs. Thus, new horizons will be opened in reconstructing the geologic history of petroliferous basins.

3. Absolute redox logs, on which the Eh potential will be referred to the standard hydrogen electrode will be available, providing SP-like logs with a zero reference. Comparing and correlating redox logs from neighboring wells on an absolute basis thus will materialize. This had never been possible with the conventional SP curve.

4. Redox logs will provide ways to map the areal extent of source beds and the crude oils to be expected from them and to evaluate the proximity to the various crude oil accumulations.

References

1. Amstutz, G. C. et al., *8th Symposium on Developments in Sedimentology,* Amsterdam: Elsevier Publishing Company, 1967, p. 466.

2. Baas Becking, L. G. M. et al., "Limits of the Natural Environment in Terms of pH and Oxidation-Reduction Poten-

tials," *The Journal of Geology,* 68, No. 3, (May, 1960), pp. 243-284.

   3. Chilingar, G. W., *8th Symposium,* p. 240.

   4. Colombo, U., G. Salimbeni, G. Sironi and I. Veneziani, "Differential Electric Log," *Geophysical Prospecting,* 7, No. 1, (1959), pp. 91-118.

   5. Dapples, E. C., *8th Symposium,* p. 9.

   6. _____ , *8th Symposium,* p. 101.

   7. Fairbridge, R. W., *8th Symposium,* p. 32.

   8. _____ , *8th Symposium,* p. 26.

   9. Garrels, R. M., *Mineral Equilibria,* New York: Harper Brothers Publishers, 1960, 1st ed., p. 254.

   10. Germanov, A. I. et al., "Investigation of the Oxidation-Reduction Potential of Ground Waters," *Geochemistry,* No. 3 (1959), pp. 322-329.

   11. Hunt, J. M., "How Gas and Oil Form and Migrate," *World Oil,* (October, 1968), 167, No. 5, pp. 140-150.

   12. Kerver, J. K. and C. L. Prokop, "When Oil or Gas Is Present, Watch Those Log Readings," *The Oil and Gas Journal,* 56, No. 50, (October 15, 1958), pp. 102-106.

   13. Krumbein, W. C. and R. M. Garrels, "Origin and Classification of Chemical Sediments in Terms of pH and Oxidation-Reduction Potentials," *The Journal of Geology,* 60, No. 1, (January, 1952), pp. 1-33.

   14. Larsen, G. and G. W. Chilingar, "Diagenesis in Sediments," *8th Symposium,* p. 551.

   15. _____ , *8th Symposium,* p. 524.

   16. Latimer, W. M., *Oxidation Potentials,* Englewood Cliffs, New Jersey: Prentice Hall, Inc., 1938, p. 392.

   17. Michaelis, L., *Oxidation-Reduction Potentials,* Philadelphia: J. B. Lippincott Co., 1930, p. 199.

   18. Pirson, S. J., "Oil Finding by Systematic Well Log Analysis," *The Log Analyst,* 6, No. 5, (January-March, 1966), pp. 4-17.

   19. Vdovykin, G. P., "Oxidation-Reduction Potential of Formation Waters of the Northwest Cis-Causcasus and of some Surface Waters," (Russian) *Petroleum Geology,* 7, No. 5, (1963), pp. 286-290.

20. Wallace, W. E., "Observations on the SP Curve in General and Offshore Problems in Particular," *8th Transaction, SPWLA* (Society of Professional Well Log Analysts), New Orleans, June, 1968.

21. Weeks, L. G., "The Gas, Oil and Sulfur Potentials of the Sea," *Ocean Industry*, 3, No. 6, (June, 1968), pp. 43-51.

22. ZoBell, C. E., "Studies on Redox Potential of Marine Sediments," *Bulletin of American Association of Petroleum Geologists*, 30, (April, 1946), pp. 477-513.

Sedimentological Studies
by Log Curve Shapes

Sedimentological observations have shown that grain size distribution, petrofabric and mineral composition distributed in cycles throughout the sedimentation sequence characterize the different shoreline processes responsible for the various sediments.

The author finds that certain sedimentation cycles may be recognized by their curve shapes on the SP and on the short normal, induction and conductivity curves. The following sand depositions are qualitatively recognized and quantitatively characterized: (1) regressive sand, (2) transgressive sand, (3) offshore sand bar, (4) channel-fill sand bar, (5) turbidite and (6) constructive delta sedimentation sequence. The hiatus in the constructive sequence may infer the destructive phase of a delta sequence.

Quantifying the intensity of regression and transgression cycles derives a mapping parameter that may be contoured for deriving its lateral projection and thereby evaluates the probable location of regressive and transgressive fossil shorelines.

Figure 2-1 classifies the types of contact between sands and shales and from SP-curve shapes.

Regressive Sedimentation Patterns

The following units, beginning at the bottom, comprise a regressive sequence (Figure 2-2, a):

1. Basal shale unit is deposited in a low energy, deepsea reduced environment. As the sedimentary units develop at shallower sea levels, the facies becomes more definitely neritic, less reduced; and the grain size becomes coarser.

2. Well-sorted sand units develop as the shoreline retreats further, generally in an oscillating cyclic fashion, causing interlaminated silts and shales. Eventually, the clean sands predominate and may become a maximum 30 feet thick. Thin fingers on the SP curve reflect the sedimentary laminations, which are gradually getting longer as the sands become cleaner toward the top of the sedimentation unit. They exhibit a horizontal parallel symmetry.

3. A topmost clean sand unit, which is a littoral sediment of high wave energy and high SP development and in which an oxidized environment (positive Eh) may develop.

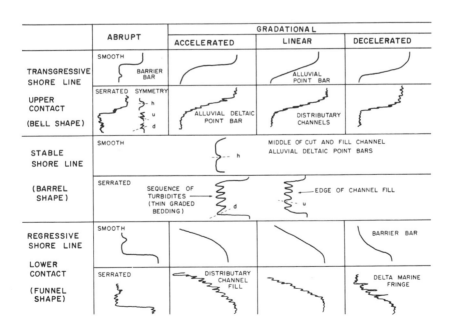

Figure 2-1. Classification of SP curve shapes in terms of sedimentary patterns.

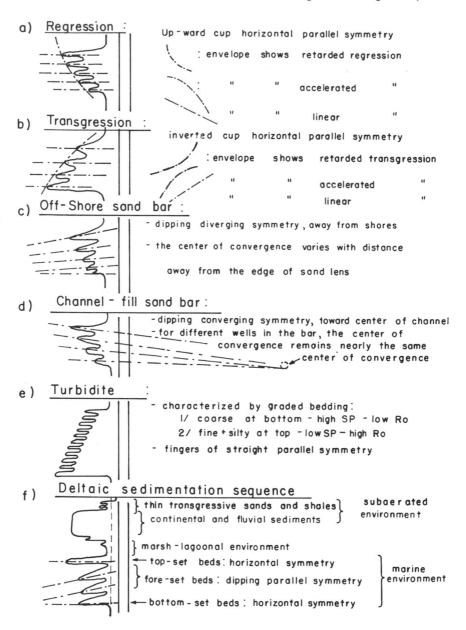

a) Regression :
              Up-ward cup horizontal parallel symmetry
                : envelope shows retarded regression
                    "   "   accelerated   "

b) Transgression :
                    "   "   linear   "
           inverted cup horizontal parallel symmetry
             : envelope shows retarded transgression
                    "   "   accelerated   "

c) Off-Shore sand bar :
                    "   "   linear   "
         - dipping diverging symmetry, away from shores
        - the center of convergence varies with distance
          away from the edge of sand lens

d) Channel - fill sand bar :
        -dipping converging symmetry, toward center of channel
        -for different wells in the bar, the center of
           convergence remains nearly the same
                  center of convergence

e) Turbidite :
        - characterized by graded bedding :
          1/ coarse at bottom - high SP - low Ro
          2/ fine + silty at top - low SP - high Ro
        - fingers of straight parallel symmetry

f) Deltaic sedimentation sequence
       } thin transgressive sands and shales } subaerated
       } continental and fluvial sediments } environment
       } marsh - lagoonal environment
       ← top-set beds : horizontal symmetry
       } fore-set beds : dipping parallel symmetry   } marine
                                  environment
       ← bottom - set beds : horizontal symmetry

Figure 2-2,a-f. Theoretical sedimentation patterns recognizable from SP curve shapes.

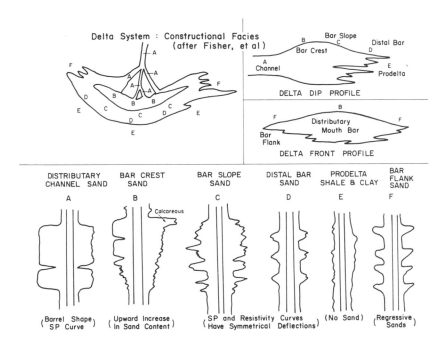

Figure 2-2,g. Theoretical sedimentation patterns recognizable from SP curve shapes.

4. The transgressive sedimentation sequence terminates the regressive sequence and brings the littoral unit in contact with a deeper environment of varying oxidation degrees, from mildly oxidized neritic shales to more reduced deep neritic marine shales.

The envelopes' general appearance to the various SP fingers characterizes the rapidity of the regressive sand deposition; accordingly:

1. A linear regression of constant sea withdrawal rate appears as a sloping straight line.

2. A decelerated regression of constantly decreasing sea withdrawal rate appears as a cusp, of which the center of curvature is toward the right or positive SP for drilling muds fresher than formation waters.

3. An accelerated regression of constantly increasing sea withdrawal rate appears as a cusp, of which the center of curvature is toward the left or negative SP.

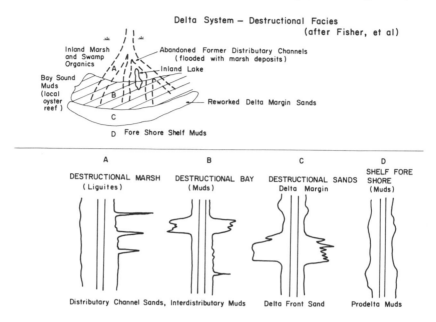

Figure 2-2,h. Theoretical sedimentation patterns recognizable from SP curve shapes.

By quantifying the observed features of the SP curves of many wells, in which the SP deflections are normalized to a constant mud resistivity, it is possible to project laterally to the strandline, which is the departure of the retreating sea or which is the departure for the next transgression. When associated with the proper transgressive environment, such shorelines will delineate stratigraphic hydrocarbon trap prospects when found up-dip or favorable uranium traps when found down-dip.

A typical regressive sand is the Olmos of Navarro age (Upper Cretaceous) in Frio County, Texas, which exhibits the characteristics described above (Figure 2-12) at various locations.

Most stratigraphic oil and gas traps in South Texas are regressive; a typical example is the Jackson trend in Jim Hogg County.

The Viking sand in Alberta, Canada, is also of Cretaceous age but not exactly at the same time as the Olmos sand. From a curve shape standpoint, however, the SP curve shapes are remarkably similar (Figure 2-3).

Figure 2-3 shows a long distance correlation Tixier and Forsythe originally presented representing a N-S cross section through the Excelsior Field, Alberta, Canada. In well 4, the Viking sand development is absent; at 3, a decelerated regression of the sand appears. Hence, it is reasonable to expect a strandline of the Viking sand to be not too far north from well 3. Further south at well 2, a constant sea withdrawal rate is indicated, whereas at well 1 the sea is regressing at an accelerated rate toward the basin's center to the west.

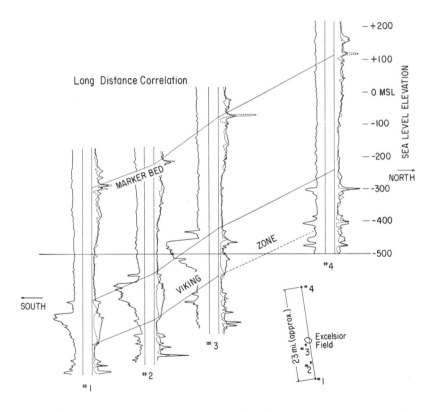

Figure 2-3. Regressive sand examples of sedimentation patterns: well 1, accelerated regression; well 2, linear regression; well 3, retarded regression; well 4, sand absent. Horizontal distance to the pinch-out of the viking sand may be determined by projecting the rate of regression retardation toward well 4. (Courtesy of M. P. Tixier, R. L. Forsythe and *The Canadian Mining and Metallurgical Bulletin*.)

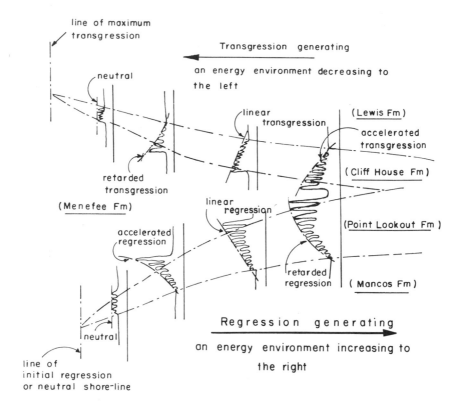

Figure 2-4. Theoretical curve shapes corresponding to various sedimentation environments simulating a northwest-southeast cross section of the Mesaverde group, San Juan Basin, New Mexico.

Transgressive Sedimentation Patterns

A transgression results from an advance of the strandline over-land resulting from a rise in sea level (Figure 2-2, b). At the transgressive series base are coarse grain sediments with good sorting followed by an abrupt grain size break. The rising sea level may be oscillatory or cyclic, thus causing sediments to grade upward into poorly sorted finer grain formations.

This sequence of sands and shales develops fingers on the SP curve, the amplitude of which decreases upward; these fingers exhibit a horizontal parallel symmetry.

The rapidity of the transgressive sea is indicated for the regressive sediments by the envelope's cusping feature to the SP fingers —yielding linear, decelerated or accelerated transgression.

By quantifying these observed features, it is again possible to calibrate the degree of transgressive acceleration and to project laterally to the strandline, which marks the position of deepest invasion overland by the transgressive sea. Often bar ridges or barrier bars similar to the Chenier ridges of the Louisiana coast are found on those strandlines.

Typical sands the transgressive process forms are the St. Peter (Illinois) and the Oriskany (Pennsylvania and New York). Figure 2-4 schematically represents the general configuration of SP curve shapes for a transgressive sequence that follows a regressive sequence for Cretaceous sediments of the Mesaverde group in the San Juan Basin, New Mexico.

The transgressive process causes only minor sedimentary volume (i.e., thin sands), when compared to regression and delta sequences, because the fundamental process is erosion and redeposition by wave action of a thin sedimentary zone as the strandline advances. The higher the transgressive sea's wave energy, the thinner the sediments. An example of transgressive sand appears in Figure 2-5 as the Tonkawa whereas the Hoover sand is regressive. Figure 2-6 also exhibits a complex pattern of transgressions and regressions.

Offshore Sand Bar

The offshore sand bar's internal structure deserves special consideration although it is formed generally during the regressive cycle (Figure 2-2, c).

Because of the generating of a sand bar by a succession of layers that wave action stacked and forming a certain angle of dip with the main depositional surface, the SP curve may show fingers exhibiting a diverging symmetry and dipping away from the strandline. The finger convergence center varies with distance from the lens edge and with its structural position in the sand bar.

Recognizing these features is often difficult because of a lack of well-defined shale breaks.

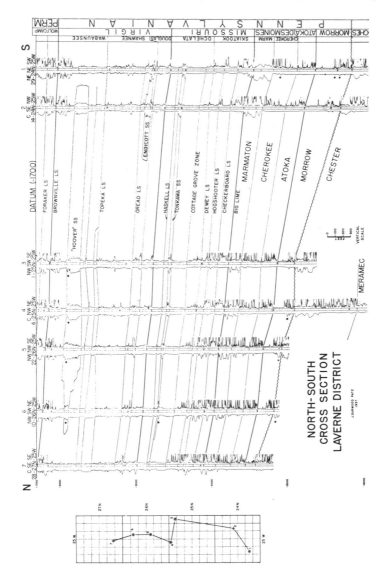

Figure 2-5. North-south cross section, Laverne district, on the north shelf of the Anadarko Basin, Oklahoma. Note, especially, the regressive feature of the Hoover sand. (Courtesy of J. Pate and AAPG.)

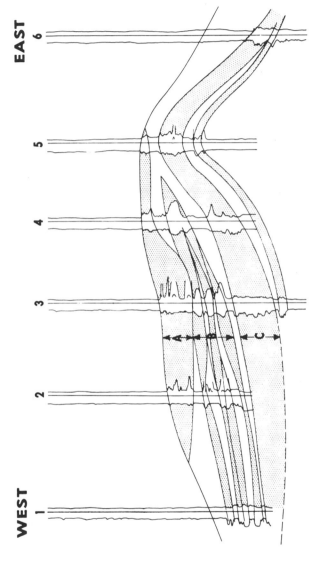

Figure 2-6. Sedimentation of stratigraphic traps illustrating complex patterns of transgression and regressions. (Courtesy of Schlumberger Well Services.)

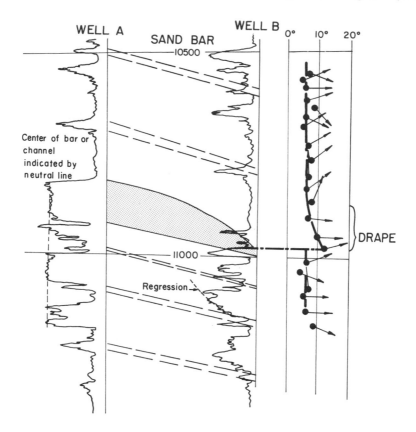

Figure 2-7. Complex sedimentation patterns in the environment of an offshore sand bar. (Courtesy of Schlumberger Well Services.)

An example of two sand bars is in Figure 2-7. The upper one is detected mostly by sediments draping over the noncompacting bar. The second bar immediately below 11,000 feet exhibits the regressive features and the diverging symmetry of the SP fingers.

Channel-Fill Sand Bar

Channel-fill sand bars (Figure 2-2, d) are generated initially by the fluvial process of river channel-cut and, then, channel-fill and valley-fill—which result in stream channels migrating laterally.

Lateral accretion of point bar deposits, which fill the entire channel from base to flood plain level gradually fills the latter.

During successive floods, different lithologies are deposited which grade upward into flood plain ooze and yield the following sequence beginning at the bottom:

1. Basal layers are poorly sorted and comprised of coarse clastic grains; large SP fingers are thus observed.

2. Very thick and massive well sorted sands of constant SP deflections overlie this.

3. Next are fine sand deposit and silt laid down horizontally by traction. Thus, fingers appear on the SP curve and decrease in magnitude upward.

4. Overlying this is a very fine sand cross-bedded with ripple marks.

5. Flood plain deposits, showing dessication cracks and a high degree of oxidation, are the uppermost horizontal layers.

The SP curve of such sediments shows finger series of various magnitudes and converging symmetry that dip toward the channel's center. If a well is located in the sand bar's center, a flat top SP curve appears; toward the edges, the SP curves look regressive. Examples of channel-fill sand bars are in Figures 2-8 and 2-9.

Turbidites

Turbidites or sediments that subsea landslides generate under gravitational influence, are recognized readily as massive silty shales of graded vertical appearance on the SP and resistivity curves (Figure 2-2, e). As the subsea landslides recur, thin sand stringers, which are extremely consistent in their sequence over extensive areas and which indicate succesive events in the turbidity current sediments, are deposited. In fact, it is difficult to explain the persistence of SP and resistivity characteristics (amplified 16 inches normal) in thick shale sections over such large areas by any other phenomenon than under sea landslide conditions during which sediments in suspension were deposited simultaneously with identical "fossil" electrical characteristics.

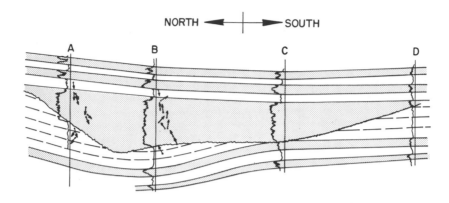

Figure 2-8. Channel-cut and fill sand. Note regressive features of wells A and C. (Courtesy of Schlumberger Well Services.)

Turbidites are "fossil" features recognizable from logs as dependable time lines. When sufficiently porous and permeable, they may also become reservoir rocks under proper conditions of later structural deformation in a favorable environment. This occurs in some California oil fields and in the Delaware basin (Delaware sand) of West Texas and Southeast New Mexico.

The Atoka formation of the Ouachita Mountains also is believed to be a turbidite.

Well-defined turbidites are unmistakable on the electric log by their linear regressive appearance on the SP curve; but this is only at a low level of deflection because of the graded-bedding nature of the nearly nondistinguishable layers' sequence.

Delta Sedimentation Sequence

Figure 2-2, f shows a delta with two periods: *constructive* and *destructive*. It is difficult, if not impossible, to reconstruct the sedimentary sequence which has been destroyed unless some vestiges of the constructive phase are well preserved in time and space.

The delta's constructive phase results from the hybrid combination of marine regression under the influx of sedimentary loads

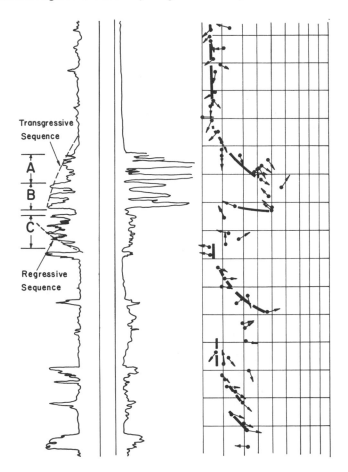

Figure 2-9. Transgressive and regressive sequences. (Courtesy of Schlumberger Well Services.)

that streams and rivers carried. Also, two main environments (marine and subareal) exist while building a delta.

During the marine accumulation, the sedimentation electric log features are typically those of a regression, initiated by a low energy sequence of sands and shales and capped by a high energy, clean clastic unit. The various facies of the constructive delta process are schematically represented in Figure 2-2g on which the various SP and resistivity curve shapes depend on well positions in the delta lobe.

When filled to sea level, partially oxidized marsh and lagoon sediments are deposited. Further sinking under the sediment load, the subareal sequence is deposited directly as distributary channel -fill sands. The sea's transgression that brings a new "pro-delta" silt and shale sequence, finally terminates the delta sequence. The various facies of the destructional phase of a delta sequence are schematically represented in Figure 2-2h, on which various SP and resistivity curve shapes depend on well positions in the delta lobe.

Two major breaks in a delta occur. The first is between the delta's marine and subareal parts; they may be recognized on the electric log by differences in redox potential, i.e., in SP shale levels. On the SP curve, an alternation in the symmetry appears. During the marine phase, the bottom-set and top-set beds exhibit a parallel horizontal symmetry in the SP fingers, whereas the foreset beds exhibit a parallel symmetry dipping toward the deep sea.

Above the marine break, the fluviatile sediments exhibit the characteristics of channel-fill sand bars.

The second major delta break occurs at the subareal phase termination when the delta cycle is resumed.

Obviously, if any part of the delta sequence is destroyed, reconstructing events will be difficult. Therefore, directional sedimentation studies using the dipmeter should prove invaluable.

Actual delta sequence examples where the interpretation is enhanced considerably by the availability of high density dipmeter results, are in Figures 2-10 and 2-11.

Sedimentation Environment Mapping

The principles of "sedimentation environment mapping" rest on geological observations that geophysicochemical differences in sediments laid in deep sea, near shore and on shore (lagoonal, neritic, marsh swamps, back swamp, shoreline, flood plain, etc.) exist.

Geologists interested in Gulf Coast sedimentation and its relation to oil generation, migration and accumulation have observed that most oil fields in stratigraphic traps are associated with maximum interfingering areas between (marine) near shore and nearly continental (deltaic) sediments. This condition is likely to generate lenticular bodies of clean sands suitable as traps that parallel the shorelines.

However, this is still considered insufficient to warrant oil accumulations. These accumulations appear when the traps are overlain by sediments that have been laid in a shallow neritic environment (oxidized flood plain sediments, Louisiana type back-swamps, salt water swamps and marshes, Bahamian carbonate platforms, mangrove saltwater swamps), which favors organic sediments' accumulating and developing.

This is believed to be a favorable environment for generating oil because in such shallow (neritic) waters, a rich vegetal and animal organic life developed; this organic matter which formed decay products (through partial oxidization) yielded oil from this source material. Subsequently, the (neritic) sediments were buried under later sediments; then, the overburden's weight compacted them, and the fluids contained therein were expelled. Thus, the oil migration and accumulation process began. The fluids' motion yielding oil accumulation was accordingly predominently downward from the overlying (neritic) sediments. Geologists have shown statistically that the downward oil accumulation from source beds was the most common and most probable process.

Favorable areas for finding oil are where closed sand lenses are developed and overlain by a (neritic) environment in close proximity.

The best possibilities for finding oil seem to be in the regressive strandline traps because the lagoonal or neritic environment moves over the prepared reservoir rock with its oil source in highly disseminated droplets. Contrarily, the reduced marine shales move over a possible transgressive strandline trap, generally carrying little oil subject to future migration.

It seems extremely important to establish the redox status of the sediments in close proximity to stratigraphic traps to determine whether a favorable environment for the oil generation existed. Properly run electric logs help map favorable redox source bed environments.

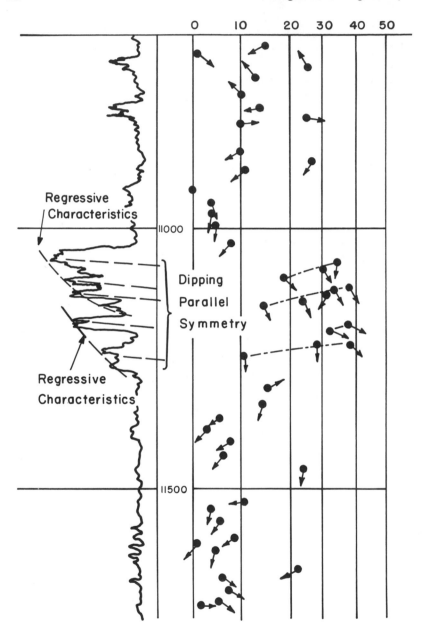

Figure 2-10. Regressive sedimentation patterns associated with foreset beds in a delta sequence. (Courtesy of Schlumberger Well Services.)

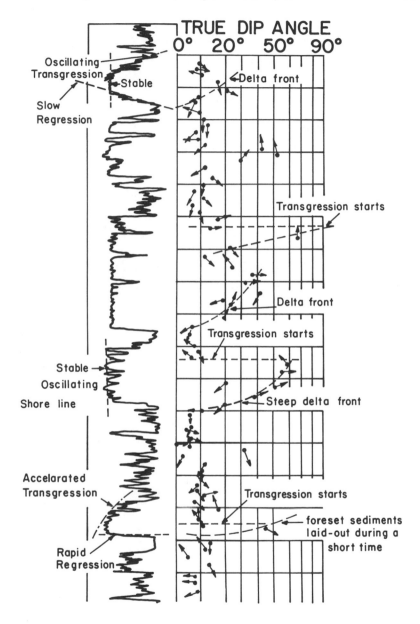

Figure 2-11. Complex sedimentation patterns that may be unravelled by combining SP curve shape and dipmeter studies. (Courtesy of Schlumberger Well Services.)

The redox log which directly measures the Eh potential of the sediments (or related potential) is, of course, a major step forward in mapping the favorable environment for oil, gas and other mineral generation, migration and accumulation.

Mapping Problem 2
Using Regressive Curve Shape Patterns to Project Laterally to a Sand Pinch-out

The prospect location (Figure 2-13) is on the northern part of the Cretaceous Shelf in Frio County, Texas, west of San Antonio. The objective sand is the Olmos B sand below the Lituola taylorensis unconformity. This sand exhibits a typical regressive pattern on the SP curve and also varies in thickness (as the well log cross section in Figure 2-12 shows). If one could find the initial regressive shoreline location and its present position on a structure, a favorable oil and gas entrapment situation could be developed.

To facilitate this, it is necessary to make the following maps: (1) a structure map on top of the Olmos B sand. (2) an isopach map of the Olmos B sand and its lateral projection to the pinch-out by projecting the thickness laterally, using a constant spacing for the isopach lines; and (3) a map of the reciprocal of the linear SP regression line in foot per millivolt, contouring and extrapolating the values to the strandline of zero foot per millivolt.

The reciprocal of the linear SP regression line comes from drawing an average sloping straight line through the SP curve. Total millivolt deflections are on this line between the base and the top of the sand. Sand thickness is divided by the deflection in millivolts, giving a regression index in foot per millivolt. These values are plotted on maps in Figure 2-13, contoured and projected laterally with constant spacing for the isoregression index lines.

The lateral projection of Maps 2 and 3 (Figure 2-13) should give substantially the same paleostrand line in position and trend to establish confidence in a stratigraphic trap's existence. It should, in addition, be located favorably on structure to justify the existence of a favorable drilling prospect. Map 1 indicates that

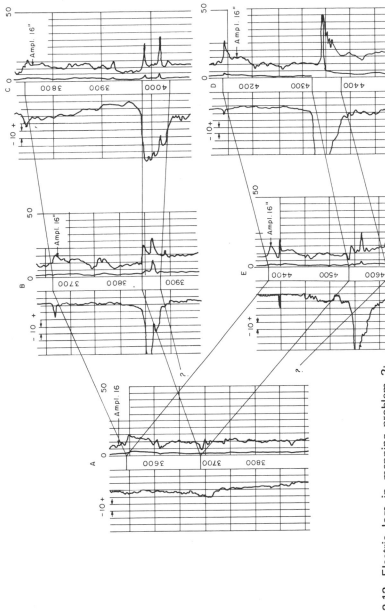

Figure 2-12. Electric logs in mapping problem 2: lateral projection to the paleo-shoreline of a regressive sand—Olmos B sand, Frio County, Texas.

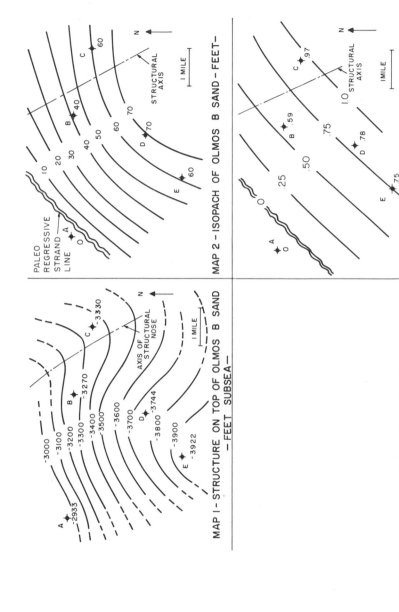

Figure 2-13. Answer to mapping problem 2 in three maps: structure, isopach and regression index.

the Olmos B sand pinch-out is on a Southeast plunging nose, and the region about two miles north of well B appears as a favorable Olmos B sand stratigraphic trap prospect.

A method similar in every respect to the above could project laterally to a transgressive strandline by determining a transgression index in foot per millivolt from the SP curves that indicate measurable transgressive features.

References

1. Arkadev, A. G. and E. M. Braverman, *Computers and Pattern Regognition,* Washington, D.C.: Thompson Book Co., 1967, p. 115.

2. Busch, D.A., "Deltas Significant in Subsurface Exploration," *World Oil,* 139, No. 7, (December, 1954), p. 95 and 140, No. 1, (January, 1955), p. 82.

3. Hollenshead, C. T. and R. L. Pritchard, *Geometry of Producing Mesaverde Sandstones, San Juan Basin,* AAPG Symposium, 1960, pp. 98-118.

4. Fons, Lloyd, *Geological Applications of Well Logs,* SPWLA Symposium, (Houston), May, 1969.

5. Passega, R., "Turbidity Currents and Petroleum Exploration," *AAPG Bulletin,* 38, (September, 1954), pp. 1871-1887.

6. Pate, J. D., "Stratigraphic Traps along North Shelf of Anadarko Basin, Oklahoma," *AAPG Bulletin,* 43, No. 1, (January, 1959), pp. 39-59

7. Saitta, S. B. and G. S. Visher, "Subsurface Study of the Southern Portion of the Bluejacket Delta," *Oklahoma Geological Society Guidebook,* 1968, pp. 52-68.

8. Scruton, P. C., "Delta Building and the Delta Sequence," in Shepard's "Recent Sediments, Northwest Gulf of Mexico," *AAPG Symposium,* Tulsa, Oklahoma, 1960, pp. 82-102.

9. Sebestyen, G.S., *Decision-making Process in Pattern Recognition,* New York: The Macmillan Company, 1962, p. 162.

10. Swann, D.H., "Rhythmic Sediments of the Mississippi Valley," *AAPG Bulletin,* 48, (May, 1964), pp. 637-658.

11. Tixier, M. P. and R. L. Forsythe, "Application of Electric Logging in Canada," *The Canadian Mining and Metall. Bulletin,* (September, 1951), pp. 358-369.

12. Visher, G. S., "Use of Vertical Profiles in Environmental Reconstruction," *AAPG Bulletin,* 49, (January, 1965), pp. 41-61.

13. West, Thomas S., "Typical Stratigraphic Traps of Jackson Trend of South Texas," *Trans. Gulf Coast Association of Geological Societies,* 13, (1963), pp. 67-78.

14. Young, R. G., "Late Cretaceous Cyclic Deposits, Book Cliffs, Eastern Utah," *AAPG Bulletin,* 41, (August, 1957), pp. 1760-1771.

Exploration for
Stratigraphic Traps

The following summary is largely from the listed literature but, more particularly, from (as yet) unpublished lectures, "Sandstones—Applied Subsurface Stratigraphy," D. A. Busch delivered as an AAPG series in applied subsurface stratigraphy.

This topic deals with subsurface stratigraphic analysis methods based on available logs from producing wells and dry holes. While useful information may also be obtained from sample cuttings, cores, fossils, etc., these notes deal only with using well logs.

Stratigraphic trap exploration basically reconstructs and restores, in three dimensions, the sedimentary conditions that existed when sand was deposited. This is accomplished by reconstructing the sand distribution patterns before their structural deformation: i.e., by restoring the depositional topography by referring (hanging) all correlatable log features to a time marker; i.e., measuring the vertical distances on the logs between the time marker and correlatable features.

The selected time markers should have been horizontally laid out at their formation. In order of decreasing quality, such markers are:

(1) Bentonite beds, recognizable on E logs by low resistivity and occasionally on RA logs by high radioactivity;

(2) Thin hard limestone beds, recognizable by their high resistivities;

(3) Coal beds, recognizable on E logs by high resistivity and on RA logs by high kicks just below the coal (probably because humates derived from coal sink to the bottom, where they absorb uranium);

(4) Silt beds (their lateral continuity is often questionable);

(5) Thin black shales (thick black shales would be dangerous to use as time markers, as their lithofacies may transgress time lines– i.e., Chattanooga shale of Upper Devonian age in Tennessee and of Lower Mississippian age in Ohio).

The differential isochronal distances so obtained are recorded on the maps, and their values are contoured to elicit the appropriate sedimentation features. The contouring must be made knowing the expected and probable geologic features that are likely to be present rather than by using mere mechanical plotters, such as associated with modern computers. The number of data points are always insufficient in achieving a complete definition of the expected geologic features, and the information the available wells provide must be supplemented with liberal doses of geologic experience, know-how and imagination.

Once the sedimentary paleosand distributions have been reconstructed, one may restore the structural deformation from topographic subsea contours on the marker bed which depict tectonic features due to the area's uplifting, tilting or rotating.

Recognizable Geologic Features on Logs

Unconformities – at times on the SP and resistivity curves; most often by known bed sequences disappearing.

Facies – nearshore (oxidized environment);
shelf (partially oxidized environment);
neritic (interbedded sands and shales, neutral environment);
abyssal (shale, reduced environment);
bathyal (shale, highly reduced environment).

Thus, facies forms a continuous redoxomorphic sequence from shore to deep sea sediments.

Time markers – listed above.

Sedimentation cycle of Genetic Interval of Strata (GIS) — the interval between an upper and lower time marker, or an unconformity, representing a continuous sedimentation process.

Genetic Sequence of Strata (GSS) — several successive and time continuous sedimentation cycles of GIS.

Methodology

The *first step* in finding stratigraphic traps [i.e., those of sedimentary bodies forming a closed or sealed external(Figure 3-1) morphological system resulting from the sedimentation process itself] is analyzing *genetic interval* of strata which is free of unconformities, disconformities or sedimentation breaks. These genetic intervals may be recognized by knowing the area's general geology. To be promising, such intervals must contain sand developments (generally not correlatable between existing wells), of which it is desirable to extend the lateral projections toward regions of expected better sand development and better existing structural positions.

The *second step* is confining the genetic interval between well-established reference levels such as these:

(1) A recognizable unconformity;

(2) Recognizable and widespread stratigraphic *time markers* (Figure 3-2) (listed above) which were deposited within a short geologic time interval and which were essentially parallel to sea level when laid. These generally may be recognized on electric and radiation logs and may be correlated over extensive areas. These stratigraphic markers should be of such quality as to be considered true time lines.

The *third step* is lining up, by correlated cross sections, all the logs using one of the upper markers as a horizontal line. Any vertical distance to an unconformity below will give the unconformity's depositional topography during a transgressing shore line. An isopach map of this interval also may be constructed to give an areal view of the paleotopography on which drainage patterns of paleostreams may become apparent when sufficient well control is available.

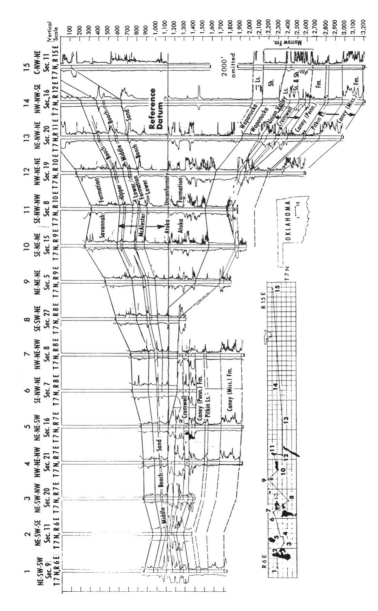

Figure 3-1. Example of electric log correlation profiling as a first step in studying regional sand development in stratigraphic trap exploration. (Courtesy of D.A. Busch and *The Oil and Gas Journal.*)

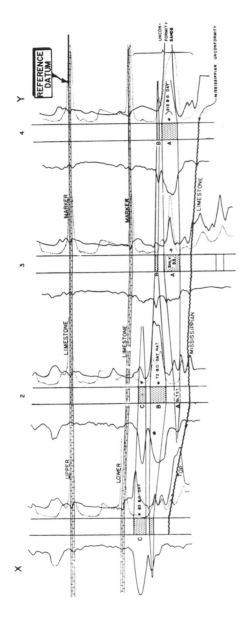

Figure 3-2. Detailed electric log profiling that outlines separate sand bench developments. (Courtesy of D.A. Busch and *The Oil and Gas Journal*.)

The cross section by logs will permit observing the regressive or transgressive nature of the sands, especially of shelf facies, and will allow projecting the expected extensions and developments of such sands laterally.

Abrupt thickening of the genetic interval is interpreted generally as a facies change and, if mostly shales are present, an abyssal facies exists. The genetic interval's isopach map will also show the rate and direction of the depositional surface's maximum slope; therefore, it will pinpoint the maximum lateral extent of certain sand developments.

The *fourth step* is studying the present structural position of the projected sand pinch-out obtained under the third step. In general, an oil accumulation in a lenticular sand results from combining two critical factors: a sealed morphological entity and a favorable structural position developed prior to the bulk migration of the fluids away from the sedimentation basin. This structural position may stem from mapping the structure on a well-established stratigraphic time marker bed. If tectonic deformation has not been too intense, the reference marker bed used under the second step may be mapped structurally. In most oil regions, this generally will be the case. When sand development and structural position confirm each other reasonably well, a favorable prospect is considered to have been developed.

A *fifth step* may prove desirable in exploring for stratigraphic traps of sand bodies—such as, beach sand, barrier bar, delta barfinger, lunate sand bar, offshore bar, channel sand, strike valley sand, etc.—especially when another stratigraphic time marker may be at the genetic interval's base. Then, it becomes possible to make a selective isopach map of the interval and to study differential compaction effects. It is well known that sand bodies are less compactible than shale beds under the overlying sediment load. Hence, thickness increase between the two stratigraphic markers—one below and one above the genetic interval—should delineate sand bodies by observing thicker intervals between markers. Again, it will be valuable to study this sand body's structural position with respect to fluid motion in the basin by constructing the structural map of a properly selected stratigraphic time marker.

Characteristics of Common Stratigraphic Traps

When the sedimentary sections are restored to their depositional topography and environment (as it existed during the genetic interval of strata and time), certain gross external geometric features recognizable on the cross sections and maps will appear. The following are the most commonly encountered sandbody developments susceptible to act as oil reservoirs:

Strike Valley Sands (Figures 3-3 and 3-4)

They are stream channels filled with sand during transgressions. These channels are between erosion-resistant cuestas carved over an unconformity surface. Such streams may be all types (consequent, subsequent, obsequent, resequent) over the old erosion surface.

Shape:

Asymmetrical biconvex
Terminate abruptly seaward
Taper toward the land

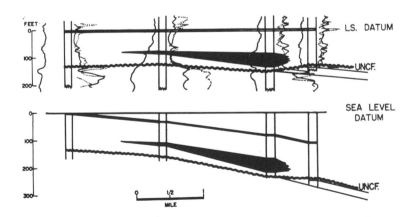

Figure 3-3. North-south profiles showing relation of strike valley sand reservoir to pre-Pennsylvanian erosion surface. (Courtesy of D.A. Busch and AAPG.)

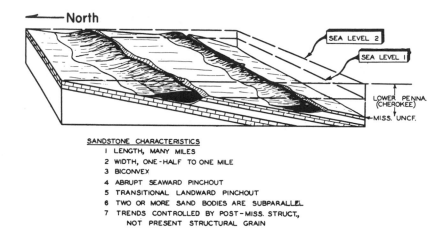

SANDSTONE CHARACTERISTICS
1 LENGTH, MANY MILES
2 WIDTH, ONE-HALF TO ONE MILE
3 BICONVEX
4 ABRUPT SEAWARD PINCHOUT
5 TRANSITIONAL LANDWARD PINCHOUT
6 TWO OR MORE SAND BODIES ARE SUBPARALLEL
7 TRENDS CONTROLLED BY POST-MISS. STRUCT.,
    NOT PRESENT STRUCTURAL GRAIN

Figure 3-4. Block diagram illustrating relation of strike valley sands to erosional escarpments developed on tilted and truncated Mississippian surface. (Courtesy of D.A. Busch and AAPG.)

Elongated paralled to escarpments
Several such sand trends subparallel to shorelines may exist
Example:
Cherokee sand, Oklahoma

Offshore Bars and Barrier Islands (Figure 3-5)

Long offshore bars are generated mostly during a regressive shoreline, but they are not excluded from formation during transgression, especially when developed over an erosion-resistant berm along shorelines.
Shape:
Plane base
Asymmetric plano-convex
Apron of sand tapering off seaward
Parallel to paleodepositional strike axes of elongated grains perpendicular to shorelines
Terminate abruptly landward
Develops thickness at the expense of overlying shale

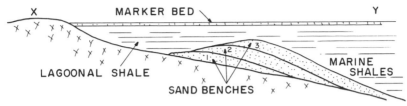

A: CROSS SECTION XY AT RIGHT ANGLE TO
PALEODEPOSITIONAL TREND

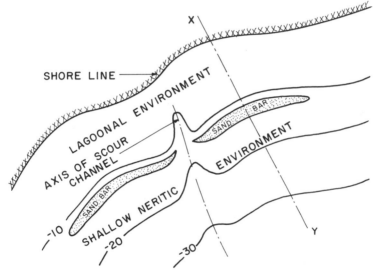

B: STRUCTURE CONTOURS ON PALEODEPOSITIONAL
SURFACE

Figure 3-5. Cross section and paleodepositional surface of an offshore sand bar or barrier island.

Possible separation by scour channels
May form, en echelon, stair-stepped benches
Differential shale compaction causes draping
Shales of different facies are on each side: logoonal (oxidized) and marine (reduced)
Example:
Shoestring-sand pools, Kansas

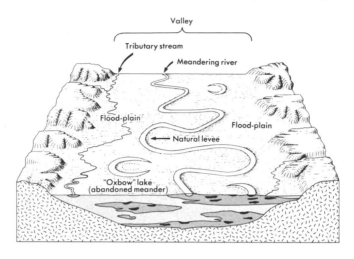

Figure 3-6. Block diagram of the valley of a meandering river showing a diagrammatic cross section. Dark gray deposits occur as irregular sandstone lenses within light gray interchannel sediments, usually shales and siltstones. Cutoff meanders may form "oxbow" lakes, where black fine-grained muds accumulate. (Courtesy of Dunbar, Rodgers and Prentice-Hall Inc.)

Channel-Fill Sand (Figure 3-6)

Channel-fill sand bars are generated by sand deposition into a stream-cut channel simultaneously during stream flow, or thereafter because of shoreline transgressions.

Shape:
Flat top
Asymmetric plano-convex
Perpendicular to paleodepositional strike
Axes of elongated grains parallel to channel axis
Develops thickness at the expense of underlying shale
Thickness increases basinward

Internal Structure:
Horizontal trough stratification (constant thickness of horizontal strata, fairly common)
Symmetrically filled with strata thickness divergence toward the axis (uncommon)
Asymmetrically filled with thickness divergence toward axis

(most common because of differential tilting and oscillations of flood plain)

Symmetrically filled without thickness divergence of the saucer-shaped strata (constant thickness strata, uncommon)

Types:

Postcompaction cut and fill (no deformation of time marker below, nor of nearby marker above)

Precompaction cut and fill (Figure 3-7) (no deformation of lower time marker, but compaction deforms upper time marker)

Concurrent deposition cut and fill (internal structure is postcompaction cut and fill with horizontal time lines)

Cut and fill cyclic subsidence (cyclic marine transgression

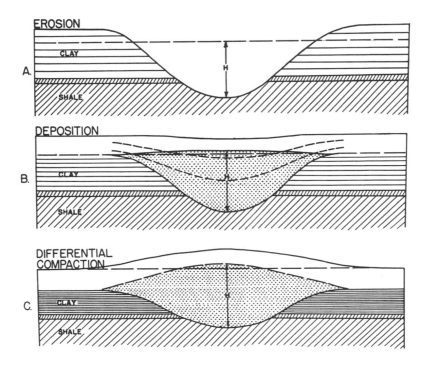

Figure 3-7. Schematic profiles of channel sand development illustrating its various development stages and the progressive development of shale compaction and draping of overlying sediments. (Courtesy of *World Oil*.)

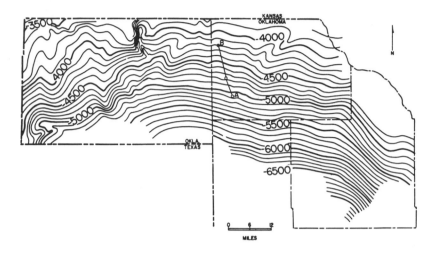

Figure 3-8. Structure of Mississippian limestone in northwestern Oklahoma. Contour interval, 100 feet. (Courtesy of D.A. Busch and AAPG.)

results in fringing beach sand on each side of the channel at each shoreline standstill; and each sand is a separate reservoir)
Example:
Bell Creek, Southeast Montana
Point-bar flood plain sand (sand deposits by slip-off toward the steep bank of river meanders; channel may fill with mud and make a lateral seal)
Example:
Coyote Creek, Wyoming
Cut and fill loading depression (delta distributary channels cut in pro-delta clay while sand accumulates in the channels with simultaneous sinking of the load; numerous such channels may be present because of differential tilting of the delta)
Example:
Lake Maracaibo reservoirs, Venezuela

Onlap of Shoreline Sands (Figures 3-8, 3-9 and 3-10)

During sea transgression over an unconformity, shoreline sands develop at each sea level standstill. Fringing beach sands develop at

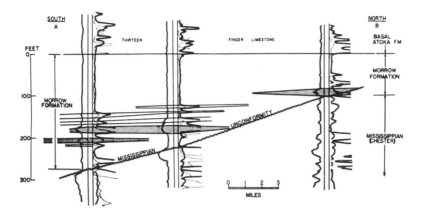

Figure 3-9. South-north cross section A-B (Fig. 3-8) showing onlap relation of several Morrow sands on the southward-dipping, truncated Mississippian. (Courtesy of D.A. Busch and AAPG.)

each shoreline standstill as the land surface subsides in cycles. During this, drainage channels that carve the unconformity surface become clogged with sand.

Shape:
Beach sands taper off rapidly toward the sea
Form stairs along unconformity surface
Example:
Morrow sands, Northwest Anadarko basin, Oklahoma

Chenier Sands

Cheniers are low, long, narrow sand ridges that develop in low swampy coastal plain areas along the coast and near the mouth of major rivers discharging on the edge of the delta plains.

Shape:
Slightly concave seaward
Slightly convex near the river
Thickness, 10 feet average
Length, up to 30 miles
Width, 600 feet average
Several are contemparaneous

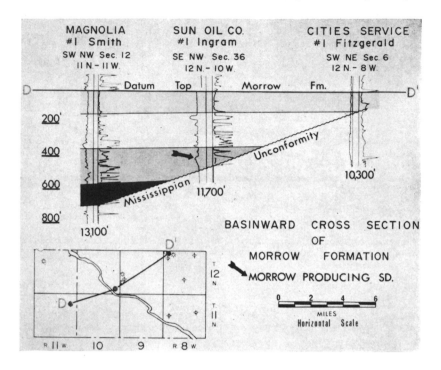

Figure 3-10. Basinward cross section of Morrow Formation, Caddo and Canadian counties, Oklahoma. (Courtesy of *World Oil.*)

Regress toward sea by incremental deposition
No separation by lagoon
Shales are of same facies on both sides
Fan out from river mouth
Example:
Chanute Pool, Kansas

Delta Sands (Figures 3-11 and 3-12)

Delta sands are deposited at a river mouth when it can no longer carry its load. The shape of the deposited sand bodies greatly depends on the load the river carries (i.e., what the fraction of the maximum load is), the type of load (in suspension or in solution) and the relative densities of the loaded river water and

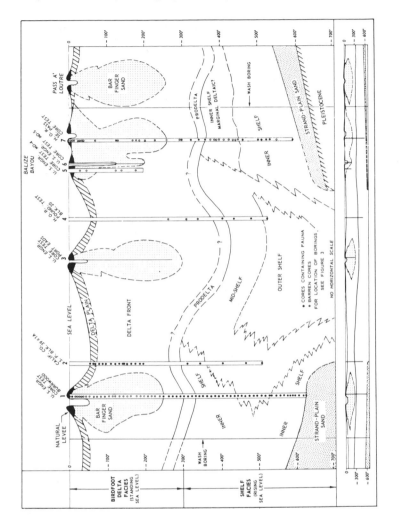

Figure 3-11. Sequence of delta-platform facies and underlying older recent shelf facies. (Courtesy of H.N. Fisk and AAPG.)

sea water into which the river discharges. Generally, the river water density is less than that of sea water, which gives rise to sediments characteristic of hypopycnal deltas.

Shape:
Sand levies present on both sides of channels
Compound lunate sand bars prograding to sea
Foreset beds in bar fingers and lunate bars are unconspicuous ($\langle$ 1° dip)
Delta occur in multiple at each shoreline standstill
Examples:
Booch sand, Oklahoma
Seligson Field, South Texas

Sand Waves and Ridges

These are staggered elongated sand bodies that develop contemporaneously from tidal wave action (*sand waves* are parallel to predominent wave crest) and tidal wave flow into bays (*sand ridges* are parallel to predominent tidal wave flow).

Examples:
Hassi-Messaoud Field, Algeria
Chester sands, Illinois

Turbidite Sands

Because of gravity, submarine rock masses may be dislodged from their support by water motion, earth tremors or plastic deformation. Subaqueous movement may thus occur as rock fall, slumping and mud flow with limited transportation distance. However, if the unconsolidated, dislodged rock mixes thoroughly with water, it forms a suspension of greater density than water which can achieve turbulent flow characteristics.

The resulting turbid mixture may travel hundreds of miles—even on gentle slopes—eventually reaching abyssal plains, where the suspended load is dumped suddenly as submarine fans that may have extensive dimensions (i.e., several hundred miles wide and long).

BRANCHING PATTERN AND THICKNESS

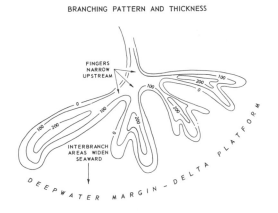

LENTICULAR CROSS SECTION

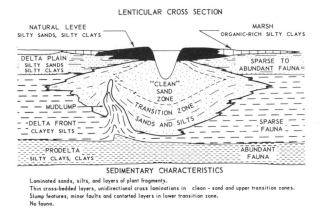

Figure 3-12. Distinguishing geometric and sedimentary characteristics of bar fingers. (Courtesy of H.N. Fisk and AAP.G.)

Example:

Grand Banks turbidite, offshore Newfoundland.

The process repeats itself over geologic time; and sequences of sediments several thousand feet thick may thus stack up even though each stratum may be only from a few inches to a few feet thick.

During the dumping of sediments, a graded bedding develops—coarse grains at the bottom, fine grains at the top.
    Example:
Delaware sand, Delaware Basin, West Texas

Mapping Problem 3
Selective Isopach Mapping

Thirty well logs were studied from an area in the Denver-
Julesburg basin, Colorado, where a new oil field (Black Jack,
Arapahoe County) was discovered in the J sand in January-
February, 1967. The discovery well is the Tiger Oil Company No.
2 UPRR Cronk, C—NE-SE, Section 9, T4S, R57W. The well
reportedly made more than 500 BPD on test. Six other producers
have been drilled, as well as a number of dry holes. The field and
well locations studied are in Figure 3-13; a cross section showing
typical logs is in Figure 3-14.
    The problem's objective is to find out from the available data
(i.e., mostly logs) if any field extension or another field in the area
and its possible extent exist. This problem may be approached by
making the following maps:
    (1) Isopach from the X-Bentonite to the top of J sand;
    (2) Isopach from the X-Bentonite to the base of J sand;
    (3) Isopach of the genetic interval of the J sand.
In addition, a quantitative evaluation of the dry holes ascertained
that they did not bypass commercial oil production. The reference
marker used for this problem is the X-Bentonite, a thin (less than
one foot thick volcanic ash bed that was spread almost simultaneously over the whole basin. The bed is most readily recognized on
the amplified short normal of the various logs by the double low
resistivity peaks together with a higher resistivity peak in between.
This occurs because of geometric distortion the electrode spacing
(AM = 16 inches) causes when it straddles the thin conductive bed.
    With induction logs, the bentonite streak appears as a single
deflection of high conductivity as the log of Well 8 shows. It may
be assumed that, during the volcanic eruption that spread out the
bentonite bed, the shallow continental shelf over which sedimenta-

TOWNSHIP. 4 S RANGE. 57 W COUNTY. ARAPAHOE STATE. COLORADO

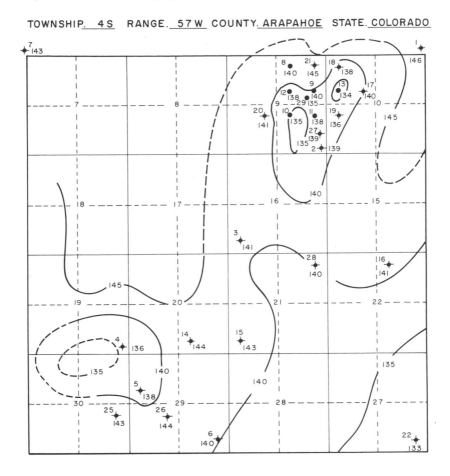

Figure 3-13. Isopach of the interval between the X-Bentonite and top of the J sand illustrating the interval's thinning over good J sand development.

tion was occurring was substantially horizontal. Therefore, the marker bed may serve as a horizontal time line. From this marker, the footage to the top of the J sand was read and the numbers were transferred to Map 1, Figure 3-15. The figures were contoured with a contour interval of 5 feet. As expected, the region of J sand development at the Black Jack Field (Sections 9 and 10, T4S, R57W) indicates a thinning of the interval because differential compaction occurred all around the field and not over the

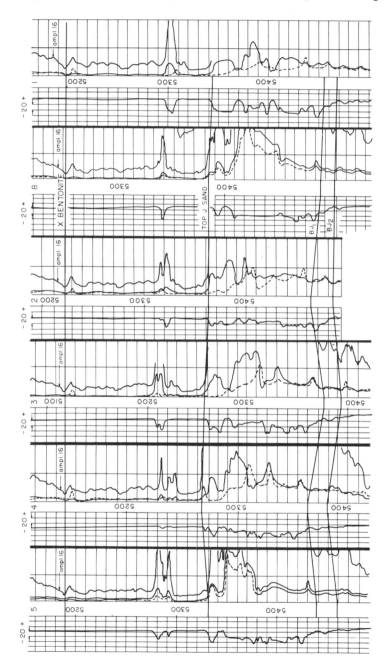

Figure 3-14. Typical logs used to map Problem 3: delineation of possible J sand stratigraphic trap, Denver-Julesburg Basin, Colorado.

TOWNSHIP. _4S_ RANGE. _57W_ COUNTY. _ARAPAHOE_ STATE. _COLORADO_

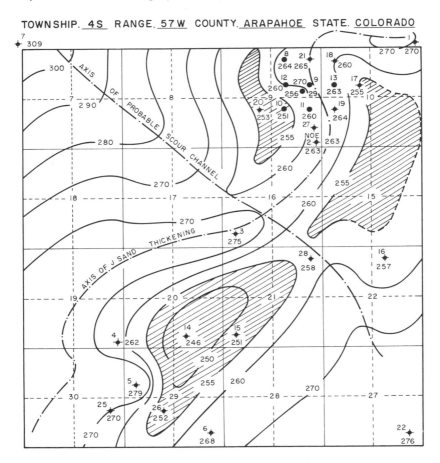

Figure 3-15. Isopach of the interval between the X-Bentonite and base of the J sand illustrating depositional topography over which the J sand was laid.

sand development, thus permitting a thicker accumulation of shale around the sand bar.

A similar thinning occurs at Wells 4 and 5, which are on the southwest trend of the old shorelines. This region could be a prospective area for discovering oil and gas accumulated under entrapment conditions similar in all respects to those at Black Jack.

To substantiate the above conclusions, a depositional trough must be under the shale thinning to warrant good sand development. The isopach map of Figure 3-15 verifies this expectation. It shows that, at the time of J sand deposition, a southwest trough extended from Black Jack through Sections 16, 17 and 19 between two log relief ridges. During the sedimentation that laid the J sand interval, this trough was filled with relatively clean sand. Some of it may have been washed away by water moving through a scour channel trending NW-SE through Sections 7 and 17.

The map of Figure 3-16 is an isopach of the genetic interval of the J sand (i.e., a map of the thickness between the top of the J sand interval and the interval's depositional surface). The latter level is particularly difficult to select, especially because all the wells do not always drill through the full J sand genetic interval. Figure 3-13 shows two possible lower markers on the cross section to which most wells were drilled and which may be reasonably identified throughout the area. The lowest one was selected for mapping.

The selection of the J sand top does not in general offer any problem, although different J sand benches may develop over a much larger area.

Map features (Figure 3-16) confirm the expectation that a thicker and cleaner J sand may have developed in Sections 19 and 30.

The selective isopach mapping technique may be refined in order to map the intervening J sand benches and study their individual area of extent and positions with respect to the low relief structures that are in the area. In order to map such structures, the X-Bentonite subsea levels would be contoured. When the well control is as dense as that in the Denver-Julesburg Basin (averaging one well per square mile) extensive detailed subsurface mapping may be resorted to by selective isopach mapping to dissect the sand's genetic interval into various benches. Field size to be so discovered, however, may not be expected on the average to be greater than the well density.

TOWNSHIP. 4S RANGE. 57W COUNTY. ARAPAHOE STATE. COLORADO

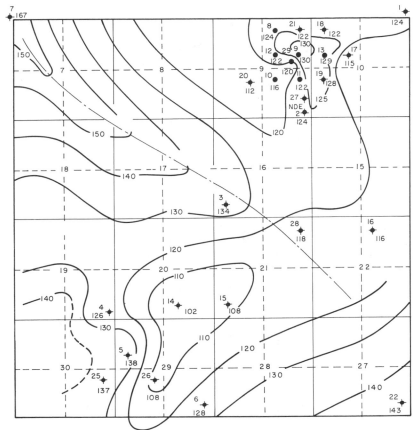

Figure 3-16. Isopach of the J sand's genetic interval illustrating the interval's thickening, where clean J sand developed.

It may be concluded that an interesting J sand prospect has been developed in Sections 19 and 20. Wells 4 and 5 reported good oil and gas shows further justifying the expectation. This conclusion would justify another test in C-NE-NW of Section 30, T4S, R57W.

References

1. Andresen, M. J., "Paleodrainage Patterns; Their Mapping from Subsurface Data and Their Paleogeographic Value," *AAPG Bulletin*, 46, No. 3, (March, 1962), pp. 398-405.

2. Busch, D. A., "Make the Most of Those Electric Logs," *The Oil and Gas Journal*, 59, No. 28, (July 10, 1961), pp. 162-167.

3. _____, "Deltas Significant in Subsurface Exploration," *World Oil*, 139, No. 7, (December, 1954), pp. 95-99 and (January, 1955), 140, No. 1, pp. 82-86.

4. _____, "Prospecting for Stratigraphic Traps," *Geometry of Sandstone Bodies*, AAPG Symposium, 1960, pp. 220-232.

5. Fisk, H. N. et al., "Sedimentary Framework of the Modern Mississippi Delta," *Journal of Sedimentary Petrology*, 24, No. 24, (1954), pp. 76-99.

6. _____, "Bar-finger Sands of Mississippi Delta," *Geometry of Sandstone Bodies*, AAPG Symposium, 1960, pp. 29-52.

7. Klinger, R. and R. Ash, "Geologic Studies Can Project Stratigraphic Trap Trends," *World Oil*, 164, No. 2, (February, 1967), pp. 65-70.

8. Martin, R., "Paleogeomorphology and Its Application to Exploration for Oil and Gas," *AAPG Bulletin*, 50, No. 10, (October, 1966), pp. 2277-2311.

9. McKee, E. D., "Primary Structures in Some Recent Sediments," *AAPG Bulletin*, 41, No. 8, (August, 1957), pp. 1704-1747.

10. Nanz, R. H., Jr., "Genesis of Oligocene Sandstone Reservoir, Seeligson Field, Jim Wells and Kleberg Counties, Texas," *AAPG Bulletin*, 38, No. 1, (January, 1954), pp. 96-117.

11. Pepper, J. F. et al., "Geology of the Bedford Shale and Berea Sandstone in the Appalachian Basin," *USGS Professional Paper 259*, 1954, p. 111.

12. Peterson, J. A. and J. C. Osmond, *Geometry of Sandstone Bodies*, AAPG Symposium, 1960.

13. Scruton, P. C., "Delta Building and the Delta Sequence" in F. P. Shepard's *Recent Sediments, Northwest Gulf of Mexico,* AAPG Symposium, 1960, pp. 82-102.

14. Swann, D. H., "Late Mississippian Rhythmic Sediments of Mississippi Valley," *AAPG Bulletin,* 48, No. 5, (May, 1964), pp. 637-658.

15. Van Siclen, D. C., "Depositional Topography—Examples and Theory," *AAPG Bulletin,* 42, No. 8, (August, 1958), pp. 1897-1913.

16. Visher, G. S., "Use of Vertical Profiles in Environmental Reconstruction," *AAPG Bulletin,* 49, No. 1, (January, 1965), pp. 41-61.

17. Wulf, G. R., "Lower Cretaceous Albian Rocks in Northern Great Plains," *AAPG Bulletin,* 46, No. 8, (August, 1962), pp. 1371-1415.

Continuous Dipmeter
as a Structural Tool

The dip angle and direction of the strata that a drill hole penetrates are of great practical interest to the oil industry. Such information is essential in studying geological structures and selecting new well locations.

Before wireline dipmeters were developed, studying oriented cores could sometimes determine the strata dip. Other methods have been tried but generally with limited success because they failed to meet the requirements for fast, yet complete and accurate measurements.

Dip determinations are usually made by correlating the logs of at least three wells not in a straight line. If correlation can be determined, homologous points of the three logs define the bedding plane. This method discounts the possibility of geological anomalies occurring between the wells. Its main practical disadvantage is that three well logs should be available; therefore, three wells must have been drilled and logged.

Since dip evaluation from correlation between several wells is complex, costly and frequently uncertain, determining the dip angle and direction of the formations has been attempted from measurements made within a single borehole.

Before describing the instruments and techniques, definitions of dip angle and direction, or azimuth, should be reviewed.

The angle which the bedding plane's "line of greatest slope" makes with a horizontal plane is the dip angle. The angle this horizontal projection of the line of greatest slope makes with the north is the dip direction, or azimuth.

Magnetic devices make all azimuth measurements. Hence, such measurements give the angles from magnetic north. These angles must then be corrected to geographic, or "true," north by adding (or subtracting) the magnetic declination where the measurements are made.

Principles of Measurements

The instrument for dip measurement uses three identical sets of electrodes, separated from each other by 120 degrees, whereby the centers of all three sets are in the same plane perpendicular to the apparatus' axis. This axis practically coincides with the hole's axis because the dipmeter tool is long and is kept centered by spring systems. In modern instruments (continuous dipmeters), spring mountings press the electrodes against the hole wall. In the earlier instruments SP and resistivity dipmeters), rigid supports maintained electrodes close to the wall and at a fixed distance from the axis.

When the instrument moves along the borehole, three curves are recorded simultaneously. Once the electrode system crosses the boundary between two formations with different electrical characteristics, the formation dip causes corresponding shifts in the curves to occur at different depths.

"Geometric elements of the dip" (angle and direction) can be derived from the values of these depth differences. To achieve this, the following additional information is necessary:

1. The orientation of the electrode system, defined by the azimuth of a particular electrode set, is designated as No. 1. This azimuth is the angle between the horizontal projection of a line perpendicular to the instrument's axis, which passes through the No. 1 electrode and magnetic north (or, when corrected for local magnetic declination, true north).

2. The drift and azimuth of the borehole is equivalent to that of the instrument, which is presumably well centered in the hole. These data are obtained by orienting systems attached respectively to the modern and earlier dipmeter systems.

3. A microcaliper system provides the hole diameter on the modern tools.

Figure 4-1 shows the correlation curves the three electrode sets recorded and their behavior when the electrodes cross the formation bedding plane at the left of the figure. The sharp break in each curve occurs as its corresponding measuring electrode crosses the pictured formation boundary. The displacement "L" between the curves electrodes I and II recorded and "M" between those electrodes I and III recorded are the basic elements for computing the formation boundary's dip.

Since the "L" and "M" distances are generally very small, a greatly expanded depth scale records dipmeters. (Five feet of recording equals 100 feet of hole or 1/24 of actual depth). At this expanded scale, the recordings are quite long.

Types of Dipmeters

The first successful wireline dipmeter, developed in the 1930s, was based on sedimentary formations, 'particularly shales,' being anisotropic. The information it supplied was limited to dip direction. The measurements were intricate and required precautions difficult to secure at the rig.

The SP dipmeter, introduced in 1942, recorded three SP curves with three electrodes distributed as in Figure 4-1. The orienting and directional device, which gave the orientation of the No. 1 electrode with respect to magnetic north and the borehole's drift and azimuth, was the photoclinometer attached directly to the electrode-supporting assembly. This SP dipmeter worked quite well in sand-shale territory, such as the Gulf Coast, where SP deflections are large and sharp.

In 1945 the resistivity dipmeter was developed to measure the dip in regions where the SP curves have less character. This instru-

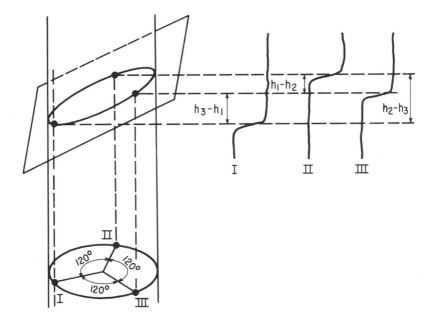

Figure 4-1. Principle of the continuous dipmeter measurements. (Courtesy of Schlumberger Well Services.)

ment records three lateral curves with 3-foot AO-spacing. The same photoclinometer again gives the orienting data. This resistivity dipmeter is more versatile than the SP dipmeter, since it can be used in hard and soft formations.

With SP and resistivity dipmeters, the curves are recorded over limited sections of hole only, since the photoclinometer shots giving the orientation data have to be taken with the dipmeter motionless in the hole. Section lengths surveyed for dip can seldom exceed 40 or 50 feet because the rotation of the instrument, as it is pulled up, cannot be accounted for over greater distances. Photoclinometer readings are always made at the bottom and top of each level and, generally at the middle.

In 1952, Schlumberger Well Surveying Corporation introduced the first continuous dipmeter—the "Continuous Dipmeter-Teleclinometer," or CDM-T. The CDM-T was originally called the

"MicroLog Continuous Dipmeter." Continuous recording of the correlation curves and the borehole drift and azimuth, as well as instrument orientation, were now available.

In 1956, the most recent version of continuous dip recording—the "Continuous Dipmeter-Poteclinometer," or CDM-P, was introduced in Oklahoma.

These continuous dipmeters, besides providing continuous correlation curves, have the advantage of using microinverse, or microfocused, devices which give correlation curves with much sharper breaks than those the SP or 3-foot Lateral devices recorded. The modern CDM-T and CDM-P have replaced the SP and resistivity dipmeters almost everywhere.

Another tool for continuous dip recording is based on correlating caliper profiles of the hole wall in three oriented planes. This tool is in limited use. In 1963, the four-arm continuous dipmeter was introduced.

Continuous Dipmeters

Both continuous dipmeters use microdevice electrode systems mounted on pads spaced 120 degrees apart around the hole wall. Spring arms apply these three pads—as they do Microlog pads—to the wall. Recording a microcaliper curve simultaneously with the dipmeter curves is also provided. The upper set of springs guides the instrument to the hole's center.

Continuous Dipmeter—Teleclinometer (CDM-T)

This instrument records, simultaneously and continuously, the three dip correlation curves, the Microcaliper curve and three orienting curves. The correlation curves are either 1 x 1-inch microinverse or the more recently developed microfocused resistivity type, depending on the equipment. The microfocused curve is similar to the one the Microlaterolog recorded. One of the orienting curves gives the azimuth of the No. 1 electrode; the other two give the east-west, north-south components of the borehole's deviation from vertical.

Continuous Dipmeter—Poteclinometer (CDM-P)

The CDM-P (Figure 4-2) differs from the CDM-T in some important respects. Its circuitry is simpler and easier to maintain; its orienting and hole drift signals are controlled by potentiometers (hence, the name—poteclinometer). It presents the drift and orientation data in terms of polar coordinates instead of rectangular coordinates. The correlation curves are all of the microfocused type, which incorporates a Microcaliper, too. The microresistivity devices are mounted on a parallelogram mechanism, which will equally displace the pads radially and apply approximately 80 pounds of equal pressure to each pad. Minimum opening is 3-1/2 inches; maximum, 17-1/2 inches. Depth of investigation of each micro device $\cong$ 2 inches.

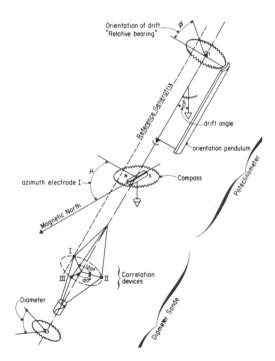

Figure 4-2. Schematic dipmeter instrumentation. (Courtesy of Schlumberger Well Services.)

This system includes these orienting curves:

(1) Its "azimuth of the No. 1 electrode" curve (a) (solid line) gives the angle from magnetic north.

(2) The "relative bearing" curve (b) (dashed line) gives the angle between the direction of the hole and the direction in which the No. 1 electrode is pointing. Specifically, it is the angle between the projections on a horizontal plane of the hole's axis and of a line perpendicular to that axis passing through the No. 1 electrode. This angle defines the hole's direction with respect to the azimuth of the No. 1 electrode.

Curve (a), however, gives the azimuth of the No. 1 electrode; therefore, the difference between (a) and (b) yields the hole's azimuth. In order to avoid too many tool rotations, a swivel head is put on top to connect to the six conductor cable.

(3) Its "hole inclination" curve is where the hole's drift angle can be read directly.

These dipmeter curves are recorded similarly to those for the CDM-T, except that, in highly resistive formations, the micro-focused curves are automatically compressed to eliminate off-scale peaks.

Figure 4-2 schematically represents the continuous dipmeter tool equipped with slide wire potentiometers for quantitative measurements of the azimuth No. 1 arm, drift angle or dip angle of the borehole's axis and relative bearing or drift direction of a vertical plane passing through the well bore's axis. The last two readings are obtained by pendular masses that continuously orient wipers that slide on the potentiometer wires.

Figure 4-3 gives the seven dipmeter curves that are recorded simultaneously—three orientation curves in the first track, three correlation curves in the second track and the borehole caliper to the right. With the old SP, or resistivity, dipmeter, discrete levels had to be chosen; and, again, the length of any given level could not exceed 40 or 50 feet because undue rotation of the instrument might possibly invalidate the results.

With the continuous dipmeters, the borehole's entire uncased portion can be surveyed. This would be too expensive for deep holes and would require a considerable amount of rig time. Gen-

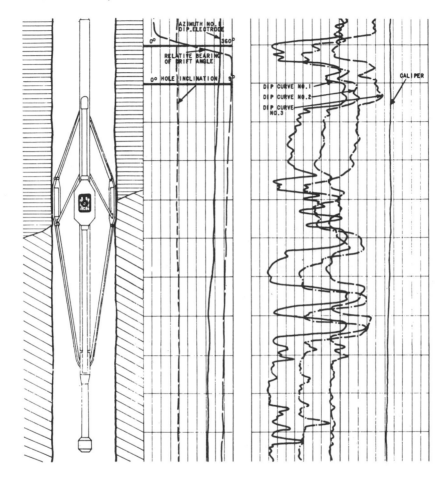

Figure 4-3. Conventional presentation of the seven continuous dipmeter curves. (Courtesy of Dresser-Atlas.)

erally, the operator knows something of the depth ranges over which he wants dip information, and the CDM survey can be run over these intervals.

Experience indicates that a dipmeter survey's usefulness depends principally on carefully selecting the well zones over which the survey is run. Obviously, numerous dip determinations at close intervals in a well can more clearly define structural and stratigraphic conditions than a few randomly chosen levels.

Each section from which a dip is to be computed should contain a number of characteristic features to be included in the over-all correlation. If the dip is measured on only one contact surface, a freak dip direction from cross-bedding might be considered as representing the true formation dip.

The sections computed must include markers which are consistent over some area and which local lensing, cross-bedding, or faulting do not greatly affect. A good way to determine this is to examine the electrical logs of nearby wells, if available, to see if the formations' electrical characteristics repeat from well to well.

The most representative dips and the most accurate and consistent correlations are provided by relatively thin beds, 2 to 10 feet thick, which contrast sharply with adjacent formations. Such zones are thin linestones or resistive sandstones interbedded with shale or thin sands or other permeable beds interbedded with shale.

Minor resistivity variations within thick sand sections or massive limestones, however, frequently give erratic dip determinations. Contacts between such sections and underlying shales are generally most reliable.

Accuracy of Results

Although it is possible to make some rough qualitative determinations in the field from the data, a specialized staff working in a central computation office has to make the final and accurate interpretation.

The drift, azimuth and orientation are read and recorded for each level or, if they vary much, for each portion of the level over which accurate correlations of the three dip curves can be made.

Very accurate measurements of the curves' relative displacements recorded by Nos. 2 and 3 electrodes to curve No. 1 are essential. Making two translucent dry prints from each original of the interval does this. These prints are then superimposed on a light table, and the curves are slid on top of each other to obtain the best possible correlation. Using two identical prints compensates for relative shrinkage or expansion due to heat from the light table, humidity, etcetera. An optical comparator measures the relative shifts more easily, quickly and accurately.

Accuracy of the measurements will vary according to the value of the dip angle and the sharpness of the "breaks" or other electrical markers on the correlation curves, and—of course—will vary with the hole diameter; because, for a given dip angle, the displacement between correlation curves will be twice as great in a 12-inch hole as it would in a 6-inch hole (assuming the same orientation of the electrodes in both cases).

Generally, for normal hole diameters of about 8 inches, dips of 10 degrees or greater can be determined with great accuracy; between 5 and 10 degrees, reliable results can be obtained when the correlation is clear and sharp and the hole inclination is not too great. Below 5 degrees, it is usually possible to obtain the dip's general direction.

Features of Continuous Dipmeters

Perhaps the most desirable feature of the continuous dipmeter is that it can be run over any length of an uncased hole section. Then, only those intervals of interest at the time need to be computed, and the original recording filed. At any later date, should dip information at other depths be required, the original can provide the necessary data.

Validity of Dipmeter Results

Correlation curves of the dipmeter survey define formation bedding planes across the well. Though these bedding planes are generally parallel to formation dip, they may, as in cross-bedding, lie at an angle with the formation dip. This fact does not belie the dipmeter's utility; on the contrary, it adds greatly to the information to be obtained. These deviations from the structural dip due to sedimentation are mostly interpretable if sufficient levels are computed to enable a close study of the pattern of dip changes.

These studies are delineating through the dipmeter such important geological features as lensing, pinchouts, and buried topography in greater detail than ever before possible.

In some massive carbonate formations, fractures and secondary porosity zones have affected dipmeter readings. These features, since they are not necessarily related to formation dip, tend to confuse the dip picture. However, the dipmeter recordings help detect and define the extent of these zones.

Causes of Inaccuracies, Errors and Reduced Quality

Mud Resistivity

Conductive muds. If Rm is larger than three ohm-meters, the quality of the correlation curves is reduced because the contact between the formation and the microresistivity electrode is greatly increased and fine variations in the profiles are smoothed out. If Rm is smaller than 0.1 ohm-meters, such as in salt muds, current leakage around the insulating pad occurs and the fine details of curve correlation disappear again. Thick mud cakes reduces correlation quality.

Oil base muds (nonconductive). Dip logging in these wells requires a knife or plow pad to scratch the nonconductive oil film and contact the formations.

Borehole Caliper Variations

Good dip calculations require a circular hole of relatively constant diameter. In washed-out sections, holes are generally elliptical or egg-shaped, and correlation curves lose a lot of their identical character and similarity required to identify correlatable stratification markers. The arms are then no longer in the same plane and calculations are incorrect. To alleviate this difficulty, the four-arm dipmeter has been used in prevalent washout areas. With the four-arm dipmeter, four solutions for the dip angle and dip direction are possible, and a most probable value at each marker then may be derived.

Highly deviated holes tend to be triangular with rounded apex into which the pads prefer to slide. This may not be a serious drawback, but the tool may get stuck by wedging.

Certain well bores exhibit a threaded surface with spiral grooves. This occurs most frequently in turbodrilled holes or in cored wells in hard formations. The effect may be recognized on the dipmeter correlation curves by a series of closely spaced peaks and valleys that are equally apart. If the peaks are taken as correlating features, they will measure constant electrode displacements, which have no significance in designating an actual bedding surface. Dip computations should thus never be made in cored intervals or in turbodrilled wells.

Lithology of Formations

The lithology of the formations drilled has an important bearing on the quality of dipmeter results.

Clastic sedimentary rocks. The most nearly universal feature of sedimentary rocks that water transport lays out is stratification (or bedding and lamination) and a degree of grain orientation, which depend on the mode of transportation and precipitation. However, such sediments as conglomerates and unconsolidated shales show no stratification.

Massive sediments. Such sediments—i.e., limestones, dolomites, chalk and reefs—show little stratification, if any, but may be highly fractured; in which case, they exhibit very erratic, apparent dips.

Accordingly, it is highly desirable to indicate on the dip-log that represents the calculated dips the correlation quality of points on the curves that were used in such calculations:

Excellent correlations = ○
Good correlations   = □
Fair correlations   = ◇
Poor correlations   = x

The grading should be based on curve similarity but also on tool rotation and hole size change. If more than eight tool rotations in 100 feet of hole are recorded, the interval should not be correlated because the orientation and resistivity markers are not reliable. In intervals where the hole size changes by 2 to 3 inches in diameter, the information's quality should be downgraded.

Techniques of Dipmeter Correlation and Computation

Correlations

Manual by shadow overlay. To avoid error by paper shrinkage, the original photographic recording film is Mylar film. It is best, then, to make one copy and lay the original film over the copy: trace 1 over trace 2, trace 1 over trace 3 with appropriate vertical displacement until a satisfactory correlation is obtained for each marker. The vertical shift and correlation quality are then noted for each.

Visual projection. It appears on a screen where the three traces are shown with appropriate vertical shifts.

Optical reader. The Seiscor instrument (Figure 4-4) is one that shows each trace in a different color: black, red and green. They are shifted optically, and the displacement is measured directly in inches and fractions thereof at the borehole scale.

Computer correlation of digitized dipmeter logs. This is a specialized work described by Moran[10] that requires large digital computing facilities.

Computations

Single marker-point computation. This is the most complete type of information and may be obtained by several techniques.

(a) *Analytic geometry* is used in solving the following equations:

$$\tan A = \frac{4}{3d} \sqrt{(h_1 - h_2)^2 + (h_1 - h_3)^2 - (h_1 - h_2)(h_1 - h_3)} \quad (4\text{-}1)$$

$$\tan B = \sqrt{3} \frac{(h_1 - h_3) - (h_1 - h_2)}{(h_1 - h_3) + (h_1 - h_2)} \quad (4\text{-}2)$$

Figure 4-4. Seiscor dipmeter log reader. (Courtesy of Seismograph Service Corporation.)

where:

A = apparent angle of the formation's maximum dip

B = azimuth of the maximum angle of the formation's dip with respect to the electrode's reading the deepest on the correlation marker.

d = hole diameter at the level of the formation being calculated

h_1, h_2, h_3 = electrode displacements.

The values of A and B have to be adjusted for the borehole deviation and tool orientation.

After rotating the axes to obtain the true azimuth (Φ) and the true dip (Θ) with respect to the earth coordinates, the following answers are obtained, letting $h_1 = 0$:

$$AN = 1.5 \, d / \sqrt{2.25d^2 + 4 \, (h_2 + h_3 - h_2 h_3)}$$

$$DN = AN \times \frac{h_2 + h_3}{1.5d}$$

$$FN = 1.15 \times AN \, (h_3 - h_2)$$

Letting: β = relative bearing angle
ϑ = hole drift or inclination
γ = azimuth = $1 - \tan^{-1} (\sin \beta / \cos \vartheta \cos \beta)$
$VN = AN \cos \vartheta + DN \sin \vartheta \cos \beta - FN \sin \vartheta \sin \beta$
$BN = - AN \cos \gamma \sin \vartheta + DN \, (\cos \gamma \cos \vartheta \cos \beta - \sin \gamma \sin \beta)$
$\qquad - FN \, (\cos \gamma \cos \vartheta \sin \beta - \sin \gamma \cos \beta)$
$EN = - AN \sin \gamma \sin \vartheta + DN \, (\sin \gamma \cos \vartheta \cos \beta + \cos \gamma \sin \beta)$
$\qquad - FN \, (\sin \gamma \cos \vartheta \sin \beta + \cos \gamma \cos \beta)$
Then: Φ = $\tan^{-1} EN/BN$ in radians
Θ = $\tan^{-1} \sqrt{1.0 - VN^2}/VN$ in radians

Digital computer programs normally solve the above equations, punching data from the seven recorded curves on IBM cards.

(b) The *Stereographic net* represents a hemisphere of a perfect globe projected on an equatorial plane. For any point on a sphere to be stereographically projected to a horizontal plane, line joining this point to the zenith point of the sphere is drawn. This connecting line's intersecting the horizontal plane through the center of the sphere locates this projected point on the equatorial plane. Figure 4-5 illustrates this stereographic projection.

Visualize a globe being rotated to place its north and south poles in a horizontal plane. Let the semicircle A A′ A″ represent a meridian on the globe's lower hemisphere. By connecting several points on the semicircle to the uppermost point of the globe (zenithal point), these points will be projected on a horizontal plane through the globe's center. An arc is formed by connecting all these points; it's the stereographic projection of the meridian line onto an equatorial plane. The complete stereographic net is constructed by projecting the meridians and the parallels of only

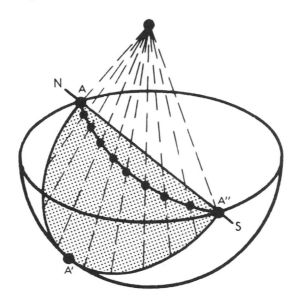

Figure 4-5. Stereographic projection. (Courtesy of Dresser-Atlas.)

the globe's lower hemisphere. Figure 4-6 illustrates a stereographic net so produced.

Running north and south and terminating at 0 to 180 degrees is a set of meridian lines spaced at 2-degree intervals and representing the great circles of the globe. Running east and west and perpendicular to the meridian lines is a set of parallel lines spaced at 2-degree intervals. This set represents the small circles of the globe.

A "small circle" is the intersection of a sphere by any plane not passing through the sphere's center. The "equator line" extends from 90 to 270 degrees through the net's middle. The circumference of the stereographic net is the "primitive" and is divided into 360 degrees by 2-degree increments. When a circle on a sphere is projected on equatorial planes, the projection's resultant curve is still an arc of a circle.

A stereographic net designed to rapidly calculate dips and strike is available from Dresser-Atlas (Form 925175) and is

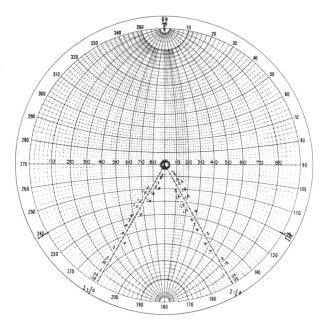

Figure 4-6. Stereographic net. (Courtesy of Dresser-Atlas.)·

provided with a plastic overlay for calculation ease. An experienced technician can do one calculation in less than two minutes.

(c) The *mechanical computer* like the Seiscor Mechanical Dip Log Computer, Figure 4-7, is a mechanical instrument on which it is possible to reconstruct borehole conditions, using the seven information items gathered from a dip log. The computer is devised so that, after having inserted the seven independent variables plus a compass correction, two dependent variables (dip direction and dip degree) can be read out directly. Utilization of swinging dip displacement rods and a floating plane greatly increases the speed at which these computations can be made.

Both 3 and 4-trace dip logs can be accommodated by using 3 and 4-arm spider assemblies, respectively. The two assemblies are completely interchangeable, and the unit is completely self-contained and portable.

Interval correlation method. In this method, an interval in the hole is selected in which all markers correlate substantially with

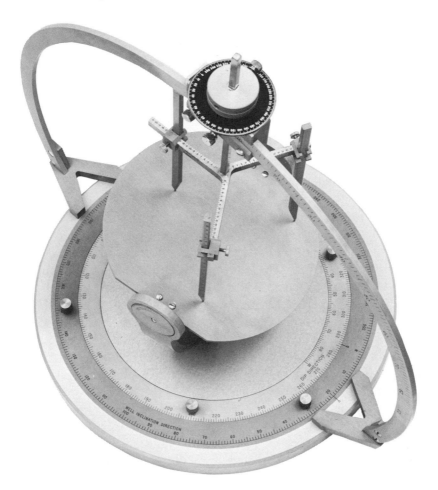

Figure 4-7. Seiscor dip log computer. (Courtesy of Seismograph Service Corporation.)

the same vertical displacements for each curve. A single computation is made for the average displacement values, and a dependable dip and dip direction are obtained for the interval. This helps to quickly detect important structural features but not sedimentation features.

High density computation method. For such a method, it is almost necessary to resort to digital computer solutions because hand calculations would demand too much time and cost. Data read manually may be punched on cards, or complete automatic computations from digitized logs may be used. Figure 4-8 shows the graphic output of such calculations.

These computed dip logs are suitable for structural and sedimentation studies.

Applying and Using Dipmeter Surveys

Determining Regional and/or Structural Dip

The primary structural dip (or regional dip if no structure is present) is the basic dip the dipmeter shows. Superimposed on this dip are dips resulting from faults, unconformities and local depositional features. These localized features generally exhibit a definite change in magnitude and/or dip direction by which they may be distinguished. In some cases, this change may show a definite trend; in other cases, the change appears in dips which are erratic in degree and direction. Primary structural dips are generally lower than those exhibited by localized features within the structure and are in a reasonably constant direction over long sections, if not the entire hole.

In finding the primary structural dip, place the dipmeter graphical chart on the wall and stand off to look at the entire survey. In 75 percent of the cases, a trend and average dip over long sections will be detected immediately. If this trend is not apparent, draw a circle divided into eight sectors and place each computed dip in the section conforming to the dip direction noted. If enough levels have been computed, even the poorest trends should be detected.

The following rules help determine those levels which indicate the structural or regional dips:

1. Rely mostly on features known to give most dependable results. Resistivity features within shale formations are considered to be most reliable. This is understandable, since shale formations are generally deposited under quieter and less localized conditions.

2. The lower dips are most likely to express the trend, since localized features generally exhibit higher dips.

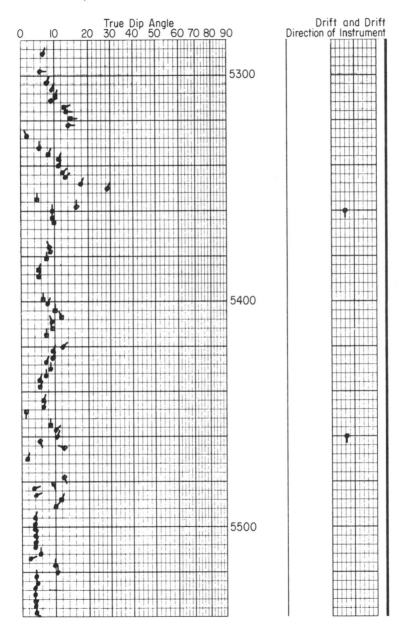

Figure 4-8. Results of high density dipmeter computations. (Courtesy of Schlumberger Well Services.)

3. If electrical logs of offset or nearby wells are available, rely more on features recognizable from well to well with little change in appearance.

The trend information obtained, along with specific dips on reservoir formations, will show structure; give an approach to its mapping; enable efficient sidetracking and offset locations; and supply thickness corrections for reservoir volume computations.

Sidetracking and Locating Offset Wells

To determine the height to be gained (or lost, if desired) on the structure in locating an offset or in sidetracking, a convenient table is given as part of the dipmeter report. This displacement in elevation is simply the tangent of the dip angle (or its component in the direction moved) multiplied by the horizontal distance between wells at the formation depth.

Determining the Reservoir's True Thickness

The dip angle may compute true thickness for reservoir volume computation. The apparent thickness from the well log multiplied by the cosine of the dip angle gives the true thickness. In certain steep dipping beds, computing true thickness may conclude that the reservoir volume is insufficient for commercial production.

Structural Interpretation

For this study, note the variations from the established trend that the dipmeter log shows. With sufficient levels computed, these variations will generally form patterns of changing dip and/ or direction. Drawing a line between each plotted point will help pick out those levels or group of levels which indicate structural features and those simply due to crossbedding. Study each zone and correlate them with available electrical and sonic logs. Note specifically any changes in resistivity or velocity associated with

irregular dips. Such changes often are due to compaction or alteration associated with faulting.

With the dip data checked and correlated with well logs and
confirmed as well as possible by other information, the structural
interpretation associates the dip changes—in particular, the pattern
of dip changes—with those geological features these changes resemble. This study utilizes one's knowledge of sedimentation and
structural geology.

Locating Faults

Faulting may appear on the dipmeter survey simply by a
sudden change in dips from one fault block to another. The most
common indication of a fault in the lower Gulf Coast is "roll-
over," or slumping of formations on the downthrown side against
the fault. This characterizes normal tension faults in incompetent
beds due to the lateral displacement. It manifests itself on the
dipmeter chart by a progressive change in dips as the borehole
approaches the fault. The schematic diagrams of Table 4-1 illustrate typical dip patterns encountered in the lower Gulf Coast and
accompanying fault conditions.

As in the schematic examples, the formation's maximum dip
may be taken as a minimum dip of the fault plane. Some general
observations based on lower Gulf Coast practice are:

1. Regional or depositional faults show more rollover than
domal faults. These faults are also called "gravity", "growth" or
"contemporaneous" faults.

2. Drag below the fault often appears but is usually limited.
This is shown by the rather rapid recovery to normal dips below
the borehole fault plane intersection.

3. No correlation has been established between the disturbed
zone's length and fault throw.

Experience in Mississippi diverges from that in the Louisiana
Gulf Coast in that drag, rather than rollover, is encountered on the
downthrown side of the fault. However, the zone of altered dips
(Text continued on page 110.)

Table 4-1 Dipmeter Interpretation Rules for Bedded and Stratified Formations

Geologic Patterns	Geologic Features	Types	Characteristics	Schematic Representation	Remarks Recognition Features
Regional Dip	Uniform Monoclinal Dip	Basinward Tilting or Isostatic Adjustment	Uniform in thickness and in lateral properties. Uniform grain orientation and tectonic strain vectors		one out of fifteen dip-vectors reflect regional dip, which is the prevalent lowest amount of dip with constant direction. Uniform grain orientation. Uniform tectonic strain vectors
Tectonic Deformation	Folding	Symmetrical Anticline	Parallel and equal dip vectors. Tectonic strain vectors parallel to axis of fold. Constant in magnitude and direction		Flank wells encounter: 1. Uniform dips 2. Uniform grain orientation 3. Uniform strain vectors
		Asymmetrical Anticline	Dips change over in magnitude and in direction. Tectonic strain vectors parallel to axis of fold and increasing toward it		Reversal of dips in magnitude and direction. Tectonic strain vectors increase in magnitude toward the axis of the structure
	Faulting	Normal Fault With Roll-Over (Gravity or Growth Fault)	Direction of maximum dip is toward downthrown block. Tectonic strain index vectors decrease in magnitude toward the fault		An offset in dip magnitude occurs at the fault plane. No strain vector
		Normal Fault With Drag	Direction of maximum dip is toward downthrown block. Tectonic strain vectors decrease toward the fault		Dip magnitudes reach a sharp peak at the fault plane. No strain vector

Table 4-1 (Continued)

Tectonic Deformation (Continued)				
	Faulting (Continued)	Thrust Fault	Maximum dips within the over-thrust block Tectonic strain vectors increase in magnitude toward the fault		If lateral thrusting is intense, this pattern may not be distinguished from an unconformity, except through the tectonic strain vectors
		Non-Distorting Fault	Regional dips are not disturbed Tectonic strain vectors undisturbed Grain orientation vector undistrubed		May not be recognized from dip measurement alone; it is necessary to observe displacement of marker beds Undisturbed orientation vectors
	Diapiric Intrusions	Salt Domes Gouge Zone Shear Zone Compressed Zone	Structural dips increase with depth Random dip within salt mass Tectonic strain index vectors tangent to dome Shear planes steeply dipping Tectonic strain vectors perpendicular to dome azimuth		The presence of a salt dome may be sensed at a distance and its position may be obtained by triangulation from tectonic strain indices
		Shale Domes	Structural dips increase with depth High apparent dips in shale mass that are shear planes		Remote sensing by tectonic strain indices and triangulation is possible as with salt domes
		Igneous Plugs (Laccoliths)	Structural dips increase with depth because of draping Random dips in igneous mass Regional dips below Typical serpentine plugs		Laccolithic masses may be detected by differential draping maps

Table 4-1 (Continued)

Geologic Patterns	Geologic Features	Types	Characteristics	Schematic Representation	Remarks Recognition Features
Sedimentation Patterns	With Differential Compaction and Draping	Sand Bars	Low energy environment: Sand grain oriented parallel to shorelines. High energy: Sand grains oriented perpendicular to shorelines. Draping of overlying beds: Dips increasing with depth		Dips of overlying beds increase with depth. Grain orientation prevails within the sand body
		Reefs	No bedding within reef. Random dips may be indicated. Draping of overlying beds. Back reef: small dips: 2° - 3°. Fore reef: steep dips: 5° to 35°		Dips of overlying beds increase with depth
	Heterogeneous Sediments Without Differential Compaction	Low Energy Cut and Fill (Flood Plain Meanders and Channels)	Wide or narrow festoon channels. Sand grains oriented perpendicular to channel axis. Maximum dips at base with decrease upward. Dips point toward axis of channels		Dips within sand body increase with depth. Point toward axis. Grain orientation prevails
		High Energy Cut and Fill	Deposition by torrential transport. Sand grains oriented parallel to erosion channel axis. Sedimentary laminations dip in cross-bedded fashion but are uniform in each bevelled block		Dips of constant direction and magnitude appear in bunches. Grain orientation prevails
		Delta Sequence	Repeated sequences of bottom set, foreset and top set beds within a thick sandy section. Low regional dip of bottom set bed toward deep water of sedimentary basin		Dips within sand body decrease with depth. Grain orientation parallel to shorelines

Table 4-1 (Concluded)

Sedimentation Patterns (Continued)				
	Heterogeneous Sediments (Continued)	Cross Bedded Sediments	Generally wind borne Dips higher than regional appear in bunches, dying out abruptly above and below High angle of bevelling	Dips within sand body decrease with depth Random grain orientation, if any
		Turbidites (Graded Bedding) and Slump Blocks	High energy with unidirectional transport within large thickness of sediment	Because of turbulence, no grain orientation
		Sediments Randomly Transported	Low energy environment without prevalent sand grain orientation Scrambled dip vector orientation, magnitude near regional	The unscrambling of the various transport directions may be made by statistical means by plotting on a polar diagram the frequency of occurence of dip directions within 30° sectors
	Sedimentation Interruptions	Parallel Unconformities	Abrupt shift in dominant dips and directions Upper beds deposited parallel to unconformity No variation in tectonic strain index	Abrupt shift in dominant dips Weathering below unconformity may be indicated
		Angular Unconformities	Abrupt shift in dominant dips and directions Upper and lower beds are imbricated to the unconformity surface No variation in tectonic strain index	Abrupt shift in dominant dips Weathering of underlying beds possible

due to the fault is no less a means of identifying and locating the fault plane.

Unconformities

Unconformities are on the dipmeter log when amount or direction of dip change in the subsequent deposition. Since unconformities generally spread extensively and are often associated in position with marker formations on the electric log, they are rather easy to recognize if the dip change exists.

Salt Overhangs

Formations on the flanks of salt domes, in general, dip more sharply as they approach and flatten away from the dome. Overhangs create a special case on the dipmeter survey. If we consider a well drilled adjacent, but not through, an overhang, dip angles increase with depth as the well nears the overhang; reach a maximum opposite it; and decrease as the well penetrates below it. A line drawn through the plotted dip calculations can actually be considered the mirror image of the salt face profile.

Structural Mapping

The continuous dipmeter has increasingly become an instrument for plotting structural geology. Dipmeter results may be used in contour mapping and vertical cross sections. In contour mapping it is reasonable to map only one contour above and below a computed point on the dipmeter survey. This is because the dip cannot be assumed to remain constant over a larger area. In contour mapping based on dipmeter results, the average of dips over an interval should be used rather than the dip at a specific point.

Figure 4-9 illustrates using the dipmeter in contour mapping. The two wells were drilled approximately 12 miles apart; correlation of well logs only indicated regional dip between them. Plotting the dips at several levels showed that a structure did exist.

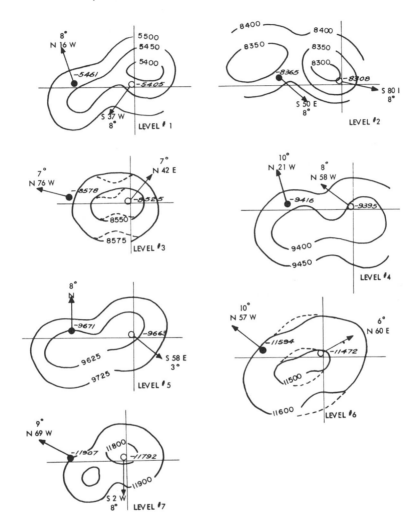

Figure 4-9. Structure mapping from dip measurements in widely separated wells. (Courtesy of Schlumberger Well Services.)

The selected levels were known correlation points on the electrical logs.

Due to the detail of the vertical dipmeter data, dipmeter results apply even more to vertical cross sections. To do this properly, it is necessary to convert each dip to its component in the plane of the section. However, if that plane lies in the trend of dips, the errors introduced are not significant.

Interpreting Directional Tectonic Trends

As a result of lateral compression from tectonic activity, a secondary petrofabric may completely destroy and replace the primary petrofabric, especially that of soft sediments (shales, marls, clays, etc.). Depending on the deformation's intensity and the sediment's competency, tectonic deformation can either partially or completely alter or erase the original depositional fabric and impose a grain orientation, which is altogether a function of stresses and strains (i.e., of deformation). Quartz, the predominant fabric element of sandstones, is very sensitive to deformation. Due to tectonic stresses in rocks, especially fine-grained rocks, regrowth of minerals and their deformation with plastic flow cause shape and fabric anisotropy. From properly calibrated and properly run dipmeter readings, the strain ellipses tectonic stresses produced may be computed in the plane of dip that correlative markers indicated. These calculations are made primarily in shale sections and at selected levels that are especially sensitive to pressure. Such levels may be recognized from other logs.

The direction of the anisotropy ellipse's main axis within the plane of dip, in addition, indicates the "degree of warping" to which the formations have been subjected. This warping is related to the intensity of fracturing by rotational strain. Warping may develop induced porosity in closely associated brittle and tight, but porous, rocks (chalk) susceptible of becoming reservoir rocks by developing permeability channels.

These exploration problems may be solved by determining *tectonic secondary petrofabric:*

1. "Proximity to a deep-seated piercement-type salt dome and direction from which the compressional stresses originated"—By using two or more wells in which continuous dipmeter logs are available, it is possible to triangulate on the region in which to expect the salt dome and possibly on the depth at which the salt

mass may be expected. Similarly, proximity to deep-seated shale masses or igneous intrusions (serpentine plugs) may be investigated.

2. "Proximity to faults that involve compression, tension stresses and strain, such as thrust and growth faults"—An unconformity is distinguished from a thrust fault in that an unconformity does not give rise to abnormal secondary petrofabric. An example in which the ambiguity was removed in interpreting the dipmeter is in Figure 4-10, where the large anisotropy at the presumed "unconformity" must be interpreted as resulting from the overlying beds' lateral thrusting. A confirmation appears in the ruffled or imbricated appearance of the lower formations from lateral overthrusting. Abnormal lengths and lineations of strain vectors parallel to the fault similarly may establish proximity to a gravity or growth fault.

3. "Proximity to fields of fracture in brittle rock"—Tracing the degree of warping with depth that has occurred in the rock sequence from tectonic deformations may ascertain this.

Results of such calculations are in Figures 4-11 and 4-12. This is from a well in Liberty County (Southeast Texas), located a few miles Northeast of the Moss Hill deep-seated salt dome. As the latter was pushed upward through formations already deposited and while others were being laid, a lateral compression was exerted in the shales toward the Northeast. Deformation vectors in magnitude and orientation are on the left in Figure 4-12.

To the right of the same figure are plotted not only the conventional dips and dip directions (particularly, in the Cockfield sand), but also the orientation of large resistivity anisotropy axes which correspond to the prevailing orientation of the elongated sand grains. Other studies have proved that the Cockfield sand in this area is an offshore bar, which has the exact orientation shown. The sand grains' orientation is, accordingly, that of the contemporaneous shorelines.

Several studies of this type have been made. They indicate that the continuous dipmeter logs can show the petrofabric (oriented internal structure) of sediments and measure *in situ* these oriented

(Text continued on page 116.)

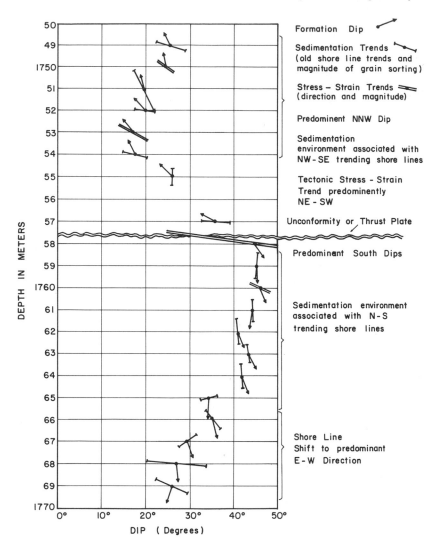

Figure 4-10. Dipmeter log interpretation including tectonic induced petro-fabric.

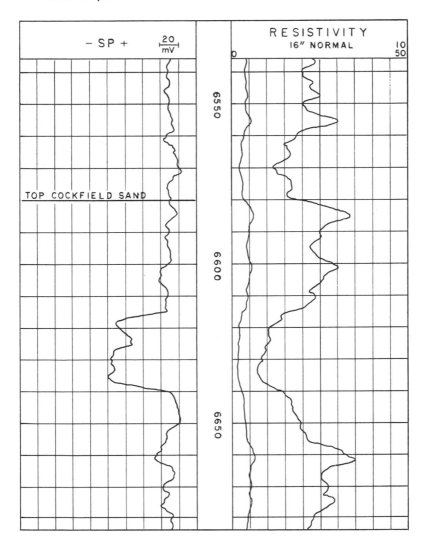

Figure 4-11. Electric log of a well in which dipmeter log interpretation helped solve tectonic deformation induced by a piercement type salt dome. (Courtesy of SPWLA.)

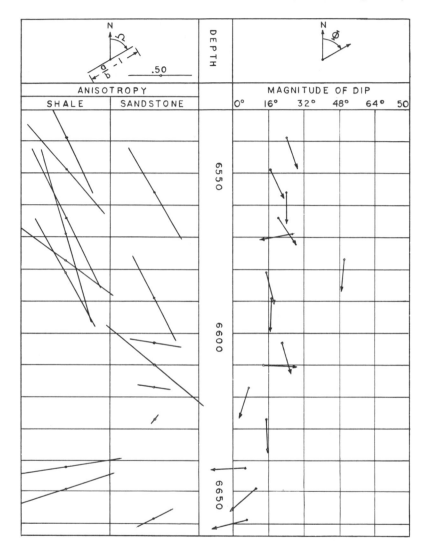

Figure 4-12. Results of dipmeter computations including tectonic strain vectors induced by a nearby piercement type salt dome. (Courtesy of SPWLA.)

structures. Its significance is that the sedimentation processes are responsible for the genesis of clastic sediments and that tectonic processes are responsible for secondary petrofabrics in a sedimentary basin.

Summary

Table 4-1 schematically diagrams geologic features and corresponding dipmeter patterns which characterize and identify them from high density dipmeter computations. These characteristics include not only the usual dip and dip directions of conventional dipmeter calculations, but also the orientation and relative intensity of the grain orientation in sands and the relative intensity and orientation of strain vectors in shales.

This comprehensive interpretation of dipmeter logs permits the unambiguous diagnosis of many geologic features otherwise unattainable by conventional means.

References

1. Boucher, F. G., A. B. Hildebrandt and H. B. Hagen, "New Dip Logging Method," *AAPG Bulletin*, 34, No. 10, (October, 1950), pp. 2007-2026.

2. Bricaud, J. M. and A. Poupon, "Le Pendage Continu en Potéclinomètre," Proceedings Fifth World Petroleum Congress, Paper No. 18, New York, June, 1959.

3. de Chambrier, P., "The Microlog Continuous Dipmeter," *Geophysics,* 18, No. 4, (October, 1953), pp. 929-951.

4. de Witte, A. J., "A Graphical Method of Dipmeter Interpretation Using the Stereo-Net," *Journal of Petroleum Technology,* 8, No. 8,(August, 1956), pp. 192-199.

5. Doll, H. G., "The S.P. Dipmeter," *AIME Journal of Petroleum Technology,* Technical Paper, 1547, (January, 1943).

6. Ferre, M. C. and R. Kinnaird, "Dipmeter Curve Comparator—An Optical Aid to the Interpretation of Dipmeter Data," *Review of Scientific Instruments,* 28, No. 9, (September, 1957).

7. Gilreath, J. A., "Interpretation of Dipmeter Surveys in Mississippi," SPWLA Symposium Transactions, Tulsa, Oklahoma, May 16-17, 1960.

8. _____ and J. J. Maricelli, "Detailed Stratigraphic Control Through Dip Computations", *AAPG Bulletin,* 48, No. 12, (December, 1964), pp. 1904-1910.

9. Grynberg, J. and M. I. Ettinger, "The Continuous Dipmeter," *Oil and Gas Journal,* 55, No. 13, (April 1, 1957), pp. 166-174 and No. 16 (April 22, 1957), pp. 129-139.

10. Moran, J. H. et al., "Automatic Computation of Dipmeter Logs Digitally Recorded on Magnetic Tapes," *AIME Journal of Petroleum Technology,* 225, (1962), pp. 771-782.

11. Prescott, B. Osborne, "Graphical Method for Calculating Dip and Strike from Continuous Dipmeters," *Oil and Gas Journal,* 52, No. 9, (March 7, 1955), pp. 118-125.

12. Schlumberger, Conrad and Marcel and H. G. Doll, "The Electromagnetic Teleclinometer and Dipmeter," *Proceedings First World Petroleum Congress,* London, 1933, pp. 424-430.

13. Stratton, E. F. and R. G. Hamilton, "Application of Dipmeter Surveys," AIME meeting, Tulsa, Oklahoma, October, 1947.

14. Thompson, J. D., "Continuous Dipmeter Survey Can Be An Important Exploration Tool," *The Oil and Gas Journal,* 59, No. 51, (December 18, 1961), pp. 128-131.

Continuous Dipmeter
as a Sedimentation Tool

A good correlation exists between petrofabric orientation and directional permeability and resistivity of rocks. Tectonic deformation of soft rocks, particularly of shales, gives rise to oriented secondary petrofabric and, thereby, to resistivity anisotropy. Accordingly, by using oriented and focused resistivities (or conductivities) recorded by the three-arm dipmeter, it is possible to derive the orientation in space of a resistivity anisotropy ellipse at selected and significant levels within the actual recorded dipmeter log. This, then, derives dominant directions of petrofabric orientation controlled by sedimentation or tectonic deformation processes.

Fabric of Sedimentary Rocks

"Fabric," as sedimentary petrologists use it, refers to the orientation of a rock's elements in space. "Apposition" or "primary" fabric is the primary orientation of a rock's elements that is developed or formed when the material is deposited. Fabrics of most sedimentary rocks are the apposition or primary type. Tectonic deformation may partially or completely modify primary fabrics in soft sediments. Examples of soft sediment deformation include flowage, sliding and disturbance by benthonic organisms.

119

A sedimentary rock's fabric element may be represented by a single crystal, detrital fragment, fossil or any component that acts as a single unit under an applied force. Quartz grains constitute the almost universal fabric elements of sandstones. Shapes of most fabric elements are like spheres, disks or rods.

Non-oriented fabric elements give rise to isotropic fabrics; whereas oriented fabric elements originate anisotropic fabrics. An anisotropic fabric results from a force field's aligning fabric elements. Force fields mainly responsible for this alignment in sedimentary rocks are the earth's gravitational field, magnetic force fields and force fields related to current flow. Such force fields have magnitude and direction and, therefore, constitute vector fields.

Spheres are not useful fabric elements for reconstructing current direction, since they cannot have a preferred orientation. They can have, however, different packing arrangements; but, in natural sands, combined mixtures of these different packing types are probably common. Furthermore, fabric elements are commonly non-spherical and unequal in size; the packing arrangement is very difficult, if not impossible, to specify in natural gravel and sand.

Disk and rod particles and particles with shapes resembling disks and rods commonly display markedly preferred orientation. Force fields mainly responsible for these particles' orientation may be the earth's gravitational field or its combination with a current force field. Particle orientation by gravity alone may occur under sedimentation in an isolated or barred basin cut off, or nearly cut off, from the main marine area. Figure 5-1 shows disks and rods oriented in a gravity field and in a combined gravity and current force field.

Sandstone and Sand Fabric

Sand fabrics can be analyzed by particulate or aggregate methods. *Particulate methods* deal with individual grains in an attempt to determine the fabric by measuring orientation of indi-

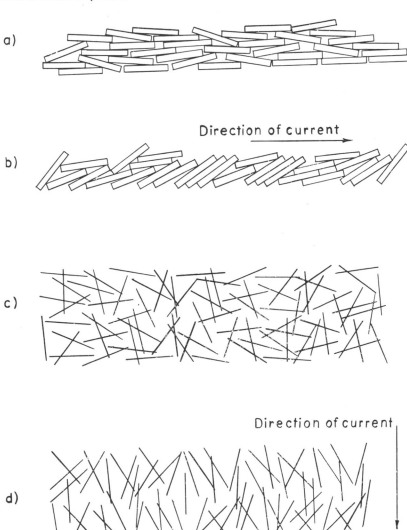

Figure 5-1. Orientation of disks and rods: (a) disks in a gravity force field, (b) disks in a combined gravity and current force field, (c) rods in a gravity force field, (d) rods in a combined gravity and current force field. (Modified from P. E. Potter and F. J. Pettijohn, courtesy of Academic Press, Inc.)

vidual grains. *Aggregate methods* of fabric determination involve measuring a bulk, or aggregate, property that depends on the orientation of many grains. Most methods of sand fabric measurement have been particulate; but, in recent years, some aggregate methods have been developed. Aggregate methods are very attractive because they integrate over a volume of sandstone that may contain 10 million to 1 billion more grains than a thin section and because they usually take up less time.

Particulate methods of sand fabric determination. Although measuring the orientation of the crystallographic "c" axes of calcite, dolomite and quartz[22] is well established and straightforward, it is not entirely satisfactory because the crystallographic "c" axis of quartz grains needs not coincide with the longest shape axis; and detrital quartz orientation depends on shape rather than crystallography. Determining the three-dimensional orientation of long axes may be made with a binocular microscope and an unmodified[26] or modified[27] multi-axis stage.

Direct orientation measurements of the apparent long axes seen in thin sections[3,11,12,23] have been made along traverses under the microscope from projecting the thin section on a screen or from photographs of the thin section. Specifying the longest apparent axis of a sand grain commonly presents a problem because quartz grains have neither perfect ellipsoidal nor rectangular shapes; and, although at least three ways have been recognized for measuring apparent long dimensions of quartz grains, a unique direction cannot be specified in some irregular grains and in those that are essentially circular.

These difficulties are of minor importance in practice—especially, if only the more elongate grains are utilized and azimuths are grouped into class intervals. However, the fabric displayed by elongate grains is only a subfabric of the sandstone. Grains counted per sample have varied widely—from 60 to 500 grains. Commonly, between 100 to 200 grains have given satisfactory results.

Aggregate methods of sand fabric analysis. An *optical method* that depends on the statistical correlation between crystallographic "c" axes and longest shape axes of quartz grains measures the varying intensity of monochromatic light passing through a thin section.[17] Data obtained with this method are in fair to good agreement with particulate fabrics, although variations in excess of 60° were obtained between the two methods on some samples.

A method of fabric analysis determines the dielectric anisotropy in rocks such as sandstones and limestones, which are relatively poor electric conductors. It is claimed that in sandstones and limestones, dielectric anisotropy correlates with fabric anisotropy. Data obtained with this method are in good to fair agreement with the particulate shape fabrics of sandstones, although some substantial deviations do occur.

An *acoustic method* determines maximum or minimum direction of acoustic anisotropy that is implied to principally depend on fabric anisotropy.[22]

A sandstone's ability to imbibe water is an anisotropic property which has been found to depend, in part, on grain fabric. Maximum imbibition of water tends to parallel the direction of the average grain elongation, and it is a basis for determining fabric anisotropy in the bedding plane.[18]

The effectiveness of a sand fabric analysis method depends on the grains involved in the analysis. Measuring a petrophysical property involves the aggregate effect of 10 million to 1 billion grains; whereas, a thin section analysis involves 60 to 500 sand grains. Consequently, an aggregate method is much more superior to a particulate.

Petrophysical Properties

In sedimentary rocks, the magnitude of petrophysical properties (such as, dielectric coefficient, electrical resistivity, fluid permeability, magnetic susceptibility, sonic transmissibility and thermal conductivity) is not the same in all directions. These properties can be represented by second rank tensor quantities.[2]

The second rank tensor associated with a quadratic form, an ellipsoid, has three principal axes: longest, intermediate and shortest. In fluid permeability, these axes correspond to the greatest,

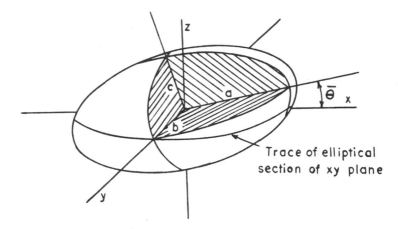

Figure 5-2. Triaxial ellipsoid of fluid permeability with a, b, c principal axes and x, y, z reference coordinate system. (Modified from P. E. Potter and F. J. Pettijohn, courtesy of Academic Press, Inc.)

intermediate and least flow direction. They are the principal axes of the medium with respect to permeability.

In Figure 5-2, consider the ellipsoid whose principal axes are "a," "b" and "c" and the rectangular coordinate system defined by "x," "y" and "z" axes. In this coordinate system, "x" axis corresponds to the direction of transport current; "y" axis parallels the depositional strike; and "z" axis is perpendicular to the "xy" plane.

Assuming the petrophysical property is fabric dependent, the following relations exist between the axes of the ellipsoid and the coordinate system: the intermediate "b" axis of the ellipsoid is colinear with "y" axis of the coordinate system; and the longest "a" axis of the ellipsoid in the "xz" symmetry plane parallels the three dimensional fabric mean. Hence, the ellipsoid is inclined to the "xy" plane at angle $\bar{\theta}$ corresponding to the fabric's average inclination angle.

Paleocurrent studies are commonly only interested in anisotropy in the "xy" plane. For an ellipsoid of given eccentricity, the anisotropy in the "xy" plane is given by the eccentricity of that plane's elliptical section.

Aggregate Petrophysical Properties of Anisotropic Fabrics

Of the sedimentary rocks' physical properties, permeability has received most attention. Several studies have demonstrated the anisotropic character of sandstone permeability—the permeability being more parallel than perpendicular to bedding. The ratio of maximum to minimum permeability in the bedding plane has varied between 1.0 and 1.5.[14] In addition, the strongest grain orientation has paralleled the direction of maximum permeability in the bedding plane.[9,10]

Good correlation has been reported between dielectric coefficient and inhomogeneities in artificial media, and fair to good correlation has been found with sandstone grain orientation. Good agreement supposedly exists between sonic anisotropy and sandstone grain orientation.

Few studies have been published on directional variations of electrical resistivity in the "xy" plane of sedimentary rocks. Nevertheless, directional electrical resistivity should correlate closely with directional permeability variations (as the Kozeny equation suggests):

$$K = \frac{1}{2t^2} \times \frac{\phi^3 \ 10^8}{S_v^2 \ (1 - \phi)^2}$$
(5-1)

where: K = absolute permeability of the porous medium in darcy

ϕ = porosity as a fraction of bulk volume

S_v = specific surface of mineral constituents, i.e., surface exposed per unit volume of rock in cm^2/cm^3

t = tortuosity coefficient of the porous network.

Since the tortuosity coefficient may be replaced by the $t^2 = F\phi$ petrophysical relation, and since the formation factor $F = R_o/R_w$, the Kozeny equation may be written as

$$K = \frac{1}{2R_o/R_w} \times \frac{\phi^2 \ 10^8}{S_v^2 \ (1 - \phi)^2}$$
(5-2)

where: R_o = resistivity of the rock fully saturated with a fluid of resistivity R_w.

This equation indicates that the directional effects of fabric on rock resistivity R_o and on permeability K are inversely proportional. Porosity and specific surface are nondirectional. This coincides with the sedimentary rocks' conducting electricity because their interconnected void spaces contain electrically conductive fluids and because the solid framework of such rocks contains minerals that generally do not conduct electricity. Experimental data on sandstones[30] have demonstrated the validity of equation

$$K_1 F_1 = K_2 F_2 \qquad\qquad (5\text{-}3)$$

where: K_1, F_1 and K_2, F_2 are the permeabilities and formation factors, respectively, measured in two different directions.

Current direction, sand fabric orientation and shape of sand bodies. Elongate sand bodies occur in all major sedimentary environments. Shoestrings, channels, bars, pods, ribbons, dendroids and belts are terms that designate such elongate sand bodies in ancient sediments. Modern equivalents of these sands include beaches or cheniers, delta and bar-finger sands, meander or point-bars in streams, offshore bars associated with the strandline.

Most elongate sand bodies are formed and shaped by those currents responsible for a preferred orientation of grains composing the sand body. Therefore, a relationship between the alignment of the long dimension of the sand body and preferred orientation of the component grains is expected.

Elongate sand bodies include two major types: those that are elongate parallel to transport direction and those that are elongate perpendicular to it. Sand bodies of fluvial origin are good examples of sand grain elongation parallel to transport direction. In fluvial deposits, substantial evidence indicates strong agreement between direction of channel elongation and sand fabric.

Beach deposits, or longshore sand bars, are good examples of sand bodies that are elongate perpendicular to the direction of the depositing current—the uprush and backwash of waves. Long axes of sand grains on the foreshore of Texas and Florida beaches have paralleled the direction of ebb and flow and, therefore, are perpendicular to the elongation of the sand body.[20] Regional studies on the Gulf Coast have further documented these findings.[4]

Without asymmetrical boundary conditions, such as on a beach, elongation of modern or ancient marine shelf sand bodies may parallel transport direction. However, discrete marine shelf sand bodies may also be elongated in a direction transverse, as well as parallel, to current direction.

Modern tidal current ridges, 25 to 100 feet high and 5 to 40 miles long, have reportedly paralleled tidal currents and, therefore, are perpendicular to shoreline. Such ridges have developed along coasts wherever tidal velocities range between 1 and 5 knots and sand is plentiful.[21]

Few data have been published on the relations between current direction and shape of turbidite sand bodies. However, like fluvial sand bodies, turbidite sand bodies should be elongate parallel to transport direction. Sand trends in modern turbidite sands off Southern California are approximately normal to strandline.[8] Turbidites may originate discrete, elongate sand bodies on deep sea fans, as well as thinner more blanket-like sand bodies in basin floors. In elongate basins, turbidite sands might be expected to be elongate parallel to the basin axis.

Oriented Petrofabric Interpretation

As a statistical expression of the resistivity anisotropy's direction and degree of preferred orientation, a simple two-dimensional scheme of vector analysis is employed.[3] The system considers the straight segments representing the anisotropy degree as vectors with direction and magnitude. Because an orientation of α^o is the same as one of $\alpha^o + 180^o$ the orientation range is 0 to 180 degrees. To express this as a periodic distribution, the angles are doubled and the vectors are summed by components. The resultant vector direction (half the angle obtained in the summation) is interpreted as the preferred orientation direction of resistivity anisotropy. The resultant vector magnitude, expressed in percent by dividing by the total readings and multiplying by 100, is interpreted as the degree of preferred orientation.

Sedimentation Interpretation

In sandstones the resistivity anisotropy orientation is perpendicular to the average preferred sand grain orientation. In turn, the average preferred sand grain orientation indicates the sand body's elongation direction on account of its dependence on the depositing current's direction.

The four common marine sand formations, which are likely to form stratigraphic traps, are listed below with the expected grain orientation. The sand body's expected trend of elongation is also indicated.

Beach sands. Long sand grain axis is aligned perpendicular to strandline. Sand body is elongated parallel to strandline.

Offshore shallow marine sands. Long sand grain axes are perpendicular to strandline. The trend of sand body elongation parallels strandline.

Deep marine sands. Preferred sand grain orientation parallels strandline. Sand body's elongated parallel to strandline.

Continental alluvial sands. Sand grain orientation and sand body elongation parallel the river valley trends.

In turbidity current deposits, the preferred grain orientation is expected to parallel the direction of current transport. In braided river deposits, grains and pebbles are oriented with their long axes perpendicular to current direction. This is because grains and pebbles are rolled along the channel bottom by the low velocity of water in the intricate network of interlacing channels.

Obviously, distinction between elongate sand bodies of strandline origin that parallel depositional strike and those of fluvial origin that are perpendicular to it is necessary to properly predict sand body elongation from directional properties. In making the needed distinction, identifying the deposition environments by other geological or biological criteria may help. The sediment's mechanical composition may suggest the kind of transporting agent. Fossils may indicate marine origin; fragmentary fossils may indicate a coastal deposit formed under shallow water. For example, marine shells at the breaker zone are always localized and fragments are handled continuously by the waves before deposition.

In attempting to define the relative position of the sediment source in a river deposit or of the strandline in a beach deposit, it may help to consider systematic changes in the characteristics of the moving load and deposited material as they are followed downstream or downbeach. These changes, which result from selective sorting or selective transportation, include a downstream decrease in average particle size, an increase in average particle sphericity and some average density changes.

Tectonic Interpretation

Tectonic deformation may partially or completely modify primary depositional fabric. Depending on the deformation intensity and sediments competency, tectonic deformation can impose a grain orientation that is the product of deformation. The less competent the sediments, the more likely the transition from a primary to a tectonic secondary fabric. Moreover, quartz, the almost universal fabric element of sandstone, is very sensitive to deformation.

Warping Index

In grain orientation resulting from tectonic deformation, the sedimentation direction deduced from the resistivity anisotropy will be apparent. Deviation of the "apparent" from the "actual" sedimentation direction will indicate the degree of tectonic deformation to which the formations have been subjected.

The degree of warping, or warping index, is the relative difference between the actual and the apparent sedimentation direction. This formula gives the warping index W:

$$W = \frac{1}{90°} \text{(TSD} - \text{ASD)} \tag{5-4}$$

where: TSD = actual sedimentation direction in degrees derived
            from regional geological information
       ASD = apparent sedimentation direction in degrees.

This parameter has most significance in brittle sediments,
where fracture porosity could develop from tectonic deformation.
Mapping the warping index may have considerable significance in
finding prospective fractured reservoirs in areas that have been
subjected to tectonic processes.

Direction of Tectonic Trends

According to De Sitter:[5]

> When rocks are distorted into any new shape, they are
> either compressed in a single direction and dilated in one or
> two directions perpendicular to the first; or compressed in
> two directions and dilated in one direction perpendicular to
> the two. The distortion may produce great folds, anticlines
> and synclines, secondary microfolds, cleavage or schistosity,
> or any other structure.

In any case, plastic yielding is always accompanied by a permanent
rearrangement of rock particles; and, therefore, a relationship is to
be expected between the electrical properties' degree of anisotropy in the bedding plane and in the lateral compression strain
and deformation resulting from lateral tectonic stresses.

The effect of lateral tectonic stresses on the anisotropy of
sedimentary rocks should be marked more on shales than on sandstones because of the former's less competent character.

Although much observational data must be gathered before
firm conclusions can be reached, field evidence suggests that in
shales the large axis of the resistivity anisotropy ellipse is perpendicular to the direction in which lateral deformation originated.

These exploration problems may be solved by determining
"tectonic secondary fabric":

*Proximity to a deep-seated piercement salt dome and direction
in which the compressional stresses originated.* From the dipmeter
data of three wells, it is possible to triangulate on the region in

which to expect the salt dome and possibly on the depth at which the salt mass may be expected. Proximity to deep-seated shale masses or igneous intrusions may similarly be investigated.

Proximity to faults where compression and tension stresses are involved (such as, thrust faults, wrench faults and growth faults). In case of uncertainty between an unconformity and a thrust fault, the ambiguity can be removed by noting that an unconformity does not give rise to abnormally oriented secondary fabric.

Determining Degree and Orientation
Of the Resistivity Anisotropy

This computation scheme determines the degree and orientation of the resistivity anisotropy:

1. By using the three resistivity values given by the three resistivity curves of a continuous dipmeter log record, calculations are made of the length of the anisotropy ellipse's semi-axes on the bedding plane and of their orientation relative to the No. 1 dipmeter electrode.

2. The degree of resistivity anisotropy is calculated from the quotient of the large semi-axis devided by the small semi-axis. In order to magnify the anisotropy effect in a graphical representation, "1" is subtracted from this quotient.

3. The orientation of the large semi-axis, as projected on a horizontal plane, is calculated with reference to the magnetic north by using the dip angle and azimuth of the bedding plane.

Figure 5-3 shows the resistivity anisotropy ellipse in the bedding plane. The large and small semi-axes are "a" and "b," respectively; R_1, R_2 and R_3 correspond respectively to resistivity values the 1, 2, 3 resistivity curves of the dipmeter log record give.

The equation of an ellipse in polar coordinates is

$$R^2 = \frac{a^2\ b^2}{a^2\ \sin^2\overline{\phi} + b^2\ \cos^2\overline{\phi}} \tag{5-5}$$

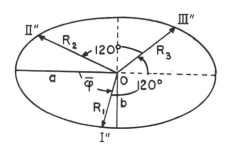

Figure 5-3. The bedding plane resistivity anisotropy ellipse created by an oriented petrofabric. (Courtesy of SPWLA.)

In determining the length of the "a" and "b" semi-axes and the value of the angle $\bar{\phi}$ that R_1 makes with the "a" semi-axis, this equation of the ellipse for the three electrodes—1", 11" and 111"—that are 120° apart is used:

$$R_1^2 = \frac{a^2 \, b^2}{a^2 \sin^2 \bar{\phi} + b^2 \cos^2 \bar{\phi}}$$

$$R_2^2 = \frac{a^2 \, b^2}{a^2 \sin^2 (\bar{\phi} + 240°) + b^2 \cos^2 (\bar{\phi} + 240°)} \qquad (5\text{-}6)$$

$$R_3^2 = \frac{a^2 \, b^2}{a^2 \sin^2 (\bar{\phi} + 120°) + b^2 \cos^2 (\bar{\phi} + 120°)}$$

Taking the reciprocals and letting $C = \dfrac{1}{R}$, rearranging and letting

$$A = \frac{1}{a^2} - \frac{1}{b^2}$$

$$B = \frac{1}{a^2} + \frac{1}{b^2}$$

the following relations result:

$$B = \frac{2}{3} (C_1^2 + C_2^2 + C_3^2)$$

$$\tan 2\bar{\phi} = \frac{\sqrt{3} (C_3^2 - C_2^2)}{2C_1^2 - C_2^2 - C_3^2}$$

$$A = \frac{2\sqrt{3} (C_3^2 - C_2^2)}{3 \sin 2\bar{\phi}}$$

The degree of anisotropy is determined from

$$\frac{a}{b} - 1 = \left(\frac{B - A}{A + B}\right)^{\frac{1}{2}} - 1 \qquad (5\text{-}7)$$

Appendix 1 gives a FORTRAN computer program, which permits the rapid calculation of dip, strike, anisotropy and anisotropy directions from dipmeter logs.

Examples

The present method of interpreting resistivity anisotropy was applied to dipmeter surveys of two wells. In each case attention was given only to that section of the dipmeter survey corresponding to a zone already known from other geological studies.

These examples are from two wells in the central part of Lake Maracaibo, Venezuela. Shaded area in Figure 5-4 corresponds to these well locations. Figures 5-5 and 5-7 trace the respective electrical logs, and Figures 5-6 and 5-8 graphically present the results of dipmeter computations.

The Santa Barbara Sand is the basal member of La Rosa Formation. This is a blanket-like formation deposited from a marine tongue or embayment extending from the northeast into the Maracaibo Basin. The preferred sand grain orientation determined from Example 1 is 40 degrees and that from Example

(Text continued on page 139)

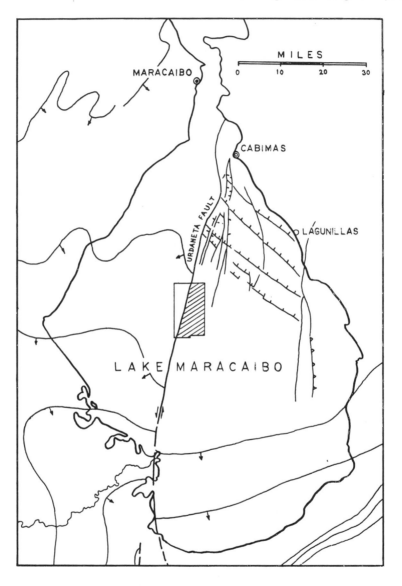

Figure 5-4. Maracaibo Lake area, Venezuela, showing locations of wells of Examples 1 and 2 with respect to the Urdaneta fault. (Courtesy of SPWLA.)

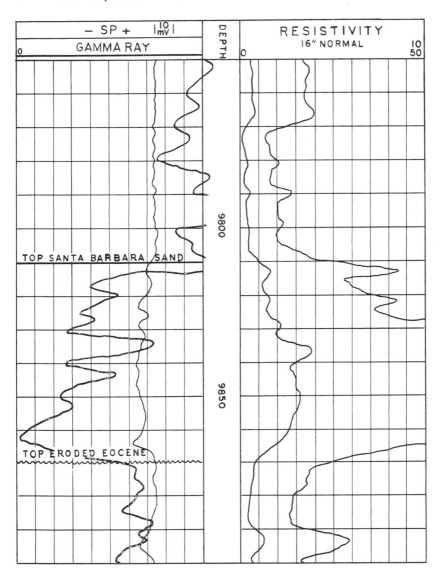

Figure 5-5. Electric log of well (Example 1) closest to the Urdaneta fault. (Courtesy of SPWLA.)

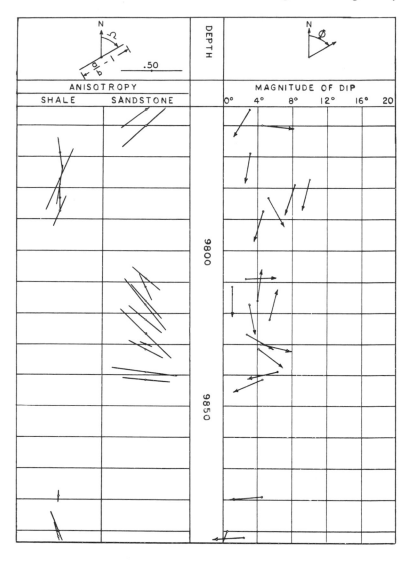

Figure 5-6. Results of Dipmeter computations for Example 1. (Courtesy of SPWLA.)

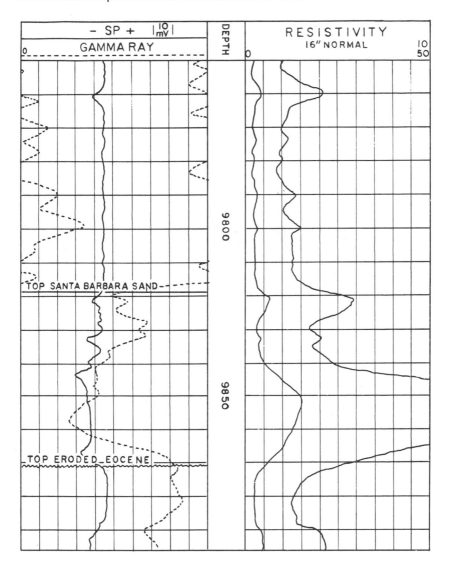

Figure 5-7. Electric log of well (Example 2) furthest away from the Urdaneta fault. (Courtesy of SPWLA.)

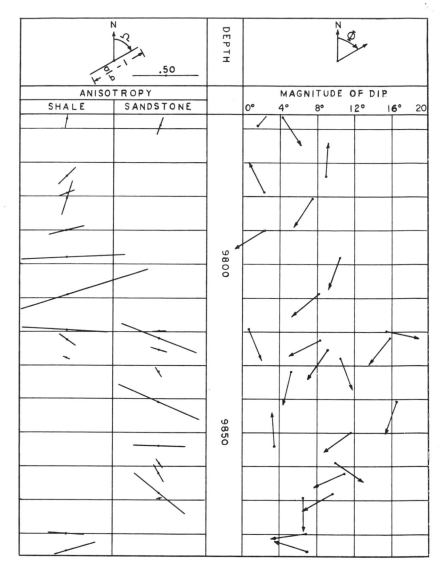

Figure 5-8. Results of Dipmeter computations for Example 1. (Courtesy of SPWLA.)

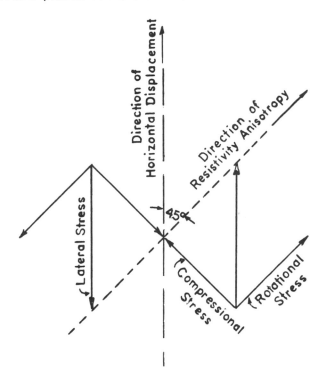

Figure 5-9. Direction of forces generated by a wrench fault and development of petrofabric anisotropy. (Courtesy of SPWLA.)

2 is 26 degrees both clockwise from north. Here again, the dipmeter's usefulness as a tool for sedimentological studies is indicated.

These two wells are very close to a wrench fault, the Urdaneta fault, which is the left strike-slip type. This fault's strike has an approximate N 15° E direction. In this fault, the lateral stress responsible for the horizontal displacement may be decomposed into two equal components—a compressional and a rotational stress (Figure 5-9).

Assuming that the resistivity anisotropy in shales is perpendicular to the lateral compressional stress, direction of the horizontal displacement should make 45 degrees with the resis-

tivity anisotropy in shales. This may account for the mean N 75°
E orientation of the resistivity anisotropy in Example 2's shales.

Directional Sedimentation Trend Studies from Conventional High Density Dipmeter Results

Directional sedimentation trend studies can predict directions
from a known well in which improved sand development may be
found once a dry hole or marginal well has been drilled and in
which the lack of, or poor productivity of, such a well may be
attributed to poor sand development. The result of such studies
may confirm or give additional information in trying to unravel
the sedimentation history of certain formations as possible
reservoirs.

Sedimentation characteristics may be derived from high
density dip computations even in areas where sedimentary units
are relatively thin and when regional dips are of low values and
low structural relief is encountered. These techniques are based
mostly on polar plots that use statistical methods to emphasize
trends and to minimize random events.

Dips computed from continuous dipmeter survey data appear
as a log of dip vectors. The log's vertical axis represents well depth
and is usually scaled with 5 inches of log representing 100 feet of
borehole. Its horizontal axis is scaled in dip magnitude with zero
dip on the left and 90 degrees on the right. The circular base of
the plotted dip vector thus indicates the depth and magnitude of
the computed dip. The azimuth of formation dip is indicated by
the polar direction of the straight line portion of the dip vector
with north toward the top of the log and parallel to the vertical
grid.

To interpret data on the dip vector log, the vectors are first
grouped according to three basic patterns. For vectors to be
grouped together to form any one of the patterns, they must
exhibit a consitent dip azimuth. The three patterns usually are
identified by these colors on the log:

Red patterns. Groups of vectors showing increasing dip magnitude with increasing depth are red. This pattern is commonly

associated with deposition over unconformities, within channels and over bar-like developments. Red patterns suggest lateral variations in the thicknesses of the lithologic units and may be associated with "drape," "creep" and "slump."

Blue patterns. Patterns of decreasing dip magnitude with increasing depth are blue. They are associated with foresetting, or current bedding, and generally indicate the sediment transport direction. They are also found just below unconformities.

Green patterns. Patterns of consistent dip magnitude and direction are green. Such patterns are prominent where structural dip magnitude is large. In areas of low structural dip, green patterns are not so prominent, and statistical methods are often needed to define structural dip.

Azimuth Frequency Diagram

An azimuth frequency diagram determines the "statistically preferred dip directions" in the geological study interval. It is a polar diagram divided in 10-degree segments with north at the top of the plot. Starting from an inner circle, segments are shaded according to the computed dip azimuths falling within the respective segments. Thus, the radial length of each shaded segment is proportional to the dip azimuths falling within that segment. In addition, azimuths from red and blue patterns in each segment are noted. With these steps, the azimuth frequency diagram tends to assume characteristic patterns as in Figure 5-10.

Bar-like Sands (barrier bars, offshore bars, tidal channels)

Bar-like sands characteristically result in two different patterns. In one, the red and blue pattern azimuths concentrate in the same general direction; in the other, the red and blue patterns fall 180 degrees apart. In each case, the red pattern concentration indicates the sand thinning direction; the blue pattern concentration indicates the direction of sediment transport during deposition; and the sand body trends at right angles to the concentrations. Grain orientation studies indicate that these sand bodies

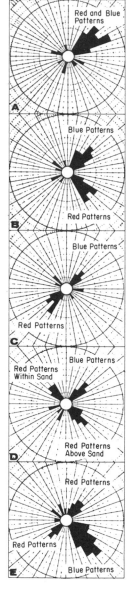

A: Bar-like sand: Red and blue patterns within sand interval point same direction. This diagram indicates sand trends N35W-S35E, thins N55E.

B: Channel or trough fill sand: Red patterns within interval point towards axis. Blue patterns point in direction of sediment transport. This diagram indicates sand trends N45E-S45W, thickens S45E, and was transported N45E.

C: Bar-like sand: Red patterns above sand, blue patterns within sand interval 180° apart. This sand trends N45W-S45E, thins S45W, and was transported N45E.

D: Channel-fill sand: Red patterns above sand 180° from red patterns within sand. Both show this sand thickens N45W, trends N45E-S45W. Blue patterns indicate transport was N45E.

E: Trough-fill sequence: Some red patterns 180° from other red patterns within fill sequence, with blue patterns at 90° to both. This well near axis of N45W-S45E trending trough. Sediments were transported S45E.

Figure 5-10. Azimuth frequency diagram used for unraveling the direction of dipmeters into meaningful sedimentation groups. (Courtesy of Schlumberger Well Services.)

are two directional deposits, one by wave motion perpendicular to shorelines and the other by longshore currents parallel to shorelines.

Trough-fill Sands (river channel deposits)

In trough-fill (or channel-fill) sands, two concentration centers 90 degrees apart are usually observed on the azimuth frequency diagram. Concentrations of red pattern azimuths point toward trough (channel) axis. Concentrations of blue pattern azimuths point toward transport. The trough (channel) trend is at right angles to the red pattern concentration. Grain orientation studies indicate that channel-fill sands are unidirectional (parallel to the trough's center).

Special Cases

While the centers of concentration usually fall in only one or two directions, three concentrations occasionally will be observed on the azimuth frequency diagram. These three-concentration diagrams characteristically have two concentrations approximately 180 degrees apart and the third at right angles.

Opposing concentrations of red pattern azimuths indicate the well is near the sand body's axis. Abundant blue patterns, at right angles, suggest a trough axis; few blue patterns suggest a ridge axis. In either case, the trend of the sand body parallels the blue pattern azimuth, or is at right angles to the red pattern azimuths.

Opposing concentrations of blue pattern azimuths may represent a series of transgressive and regressive trends within a trough. If so, the trough trends parallel to the blue pattern azimuths and its axis is indicated by the concentration of red pattern azimuths.

Using Derived Directions

Integrating dipmeter information with geologic data is an excellent way to extrapolate sand trends on isopach maps. The distance that depositional directions can be extrapolated corresponds with the vertical section which exhibits these directions. It is not uncommon for trough fill to show consistent directions in a series of red and blue patterns over several hundred feet of hole. The trend direction may be extrapolated to 640-acre offset locations with confidence when this occurs. Channel-fill, on the other hand, may show consistent directions in a series of red and blue patterns for only 25 to 50 feet. Under these conditions, 40-acre offsets may be overextending the data. Similar relationships exist for beach or bar deposits.

Trend or strike directions usually involve greater distances than the thickening direction; thus, offset locations on strike or trend are much more likely to succeed than those in the thickening direction.

Modified Schmidt Diagram

A modified Schmidt diagram, as in Figure 5-11, distinguishes between present structural and depositional dip. The diagram is a polar plot of dip azimuth and magnitude. The azimuth grid, with north at the top of the plot, divides the circle into 10-degree segments. The dip magnitude grid is formed by concentric circles corresponding to each 10 degrees of dip with zero at the perimeter and 90 degrees at the center. For the interval under study, each computed dip is plotted with respect to magnitude and direction. Contours are then drawn on equal concentrations of dip occurrences within incremental grid sectors.

In low structural dip areas, point concentrations from structure fall along the periphery of the plot. These contours are elongate with little variation in dip magnitude. The azimuth variation within the concentration of structural dips decreases as the structural dip magnitude increases. However, variation in dip magnitude remains small.

Depositional dips lead to triangle-shaped contours. The triangle base appears on or near the periphery of the plot with the apex points toward the center. Depositional dips characteristically exhibit large variations in magnitude—from zero to 40 degrees. When structural dip exceeds a few degrees, components from structural dip must be subtracted before depositional directions can be oriented properly.

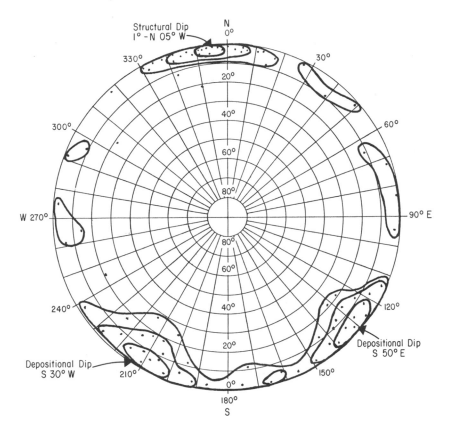

Figure 5-11. Modified Schmidt polar diagram, which discriminates between sedimentary depositional trends from regional structural trends. (Courtesy of Schlumberger Well Services.)

Additional information on procedures for stratigraphic studies may be obtained from R. L. Campbell, Jr., in "Stratigraphic Applications of Dipmeter Data in Mid-Continent," *AAPG Bulletin,* 52, No. 9, (1968), pp. 1700-1719.

(Content in this Chapter is mostly from A. R. Rodriguez' and S. J. Pirson's "The Continuous Dipmeter as a Tool for Studies in Directional Sedimentation and Directional Tectonics," Paper G, *9th SPWLA Symposium,* New Orleans, June, 1968; and a Schlumberger Technical Release.)

Mapping Problem 4
Using Continuous Dipmeter Results as a Sedimentation Tool

The log of this problem (Figure 5-12) gives high density dip-meter results obtained as graphic output from a digital computer. The well is presumably from the Gulf Coast, but its exact location is unknown.

The patterns of plotted dip vectors determine the sedimentation that has occurred. This may be done by the modified Schmidt diagram (Figure 5-11). The same diagram, however, may also plot dip readings within a 10 degree azimuth section (Figure 5-10) with five dip points per segment. This interpretation is to be made separately for two groups of beds, A and B.

Intervals A and B are selected because the dip vectors show similar patterns. Trend lines drawn through the origin of the vectors (Figure 5-13) well emphasizes this. Group A shows decreasing dips with depth in three different benches; this is normally called a blue pattern characteristic of foreset beds. Group B shows increasing dips with depth in three different benches, also—a red pattern characteristic of draping.

In order to interpret further, dip direction and dip angle of each vector are plotted in a modified Schmidt diagram. Group A's blue patterns are plotted Figure 5-14 and are separated into two groups—one trending N 40° E; the other, N 20° W. When this feature is interpreted by the rules of Figure 5-10, the blue patterns indicate the sediment transport directions. In this case, two directions predominate: N 65° E and 90° W. Group B's red patterns are plotted on Figure 5-15 and indicate dip directions or, perhaps, draping. Two dominant dip directions are N 75° E and N 75° W.

References

1. Badgley, P. C., *Structural and Tectonic Principles,* New York: Harper and Row, 1965.

2. Bouma, A. H. and A. Brouwer, "Turbidites" in "Developments in Sedimentology," Paper No. 3, Amsterdam: Elsevier Publishing Co., 1964.

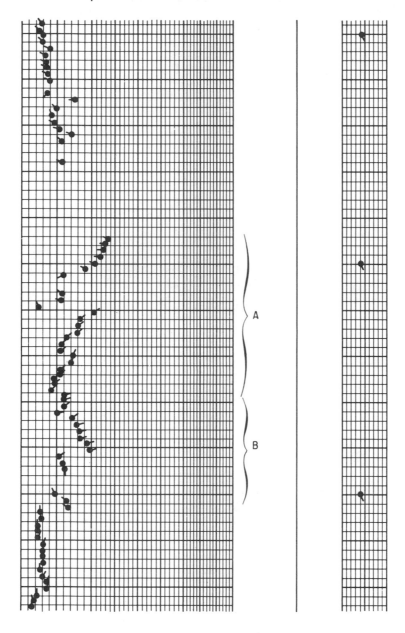

Figure 5-12. Mapping Problem 4.

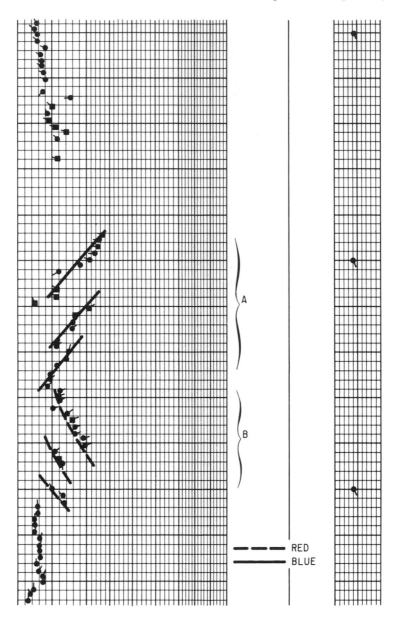

RED
BLUE

Figure 5-13. Answer to Mapping Problem 4.

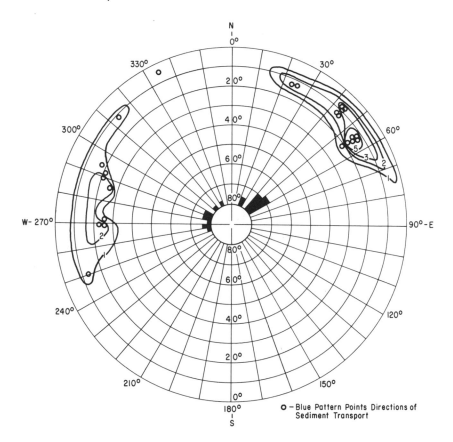

Figure 5-14. Answer to Mapping Problem 4.

3. Curray, J. R., "The Analysis of Two-Dimensional Orientation Data," *The Journal of Geology,* 64, No. 2, (1956), pp. 117-131.

4. _____ , "Dimensional Grain Orientation Studies of Recent Coastal Sands," *AAPG Bulletin,* 40, No. 10, (1956), pp. 2240-2456.

5. De Sitter, L. U., *Structural Geology,* London: McGraw-Hill Book Co., Inc., 1956, p. 552.

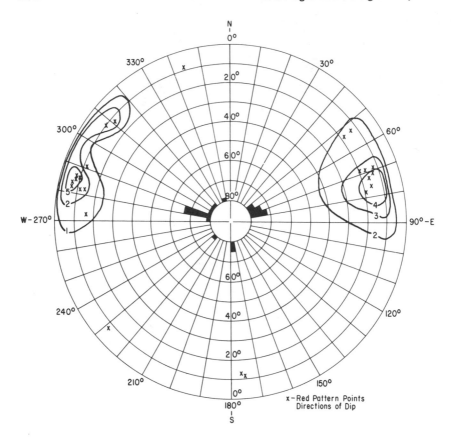

Figure 5-15. Answer to Mapping Problem 4.

6. Fraser, H. J., "Experimental Study of the Porosity and Permeability of Clastic Sediments," *The Journal of Geology,* 43, No. 8, (1935), pp. 910-1010.

7. Gilreath, J. A. and J. J. Maricelli, "Detailed Stratigraphic Control Through Dip Computations," *AAPG Bulletin,* 48, No. 12, (1964), pp. 1902-1910.

8. Gorsline, D. S. and K. O. Emery, "Turbidity-Current Deposits in San Pedro and Santa Monica Basins Off Southern

California," *Bulletin Geological Society America,* 70, (March, 1959), pp. 279-290.

9. Griffiths, J. C., "Directional Permeability and Dimensional Orientation in Bradford Sand," *Producers Monthly*, 14, No. 8, (1950), pp. 26-32.

10. _____ and M. A. Rosenfeld, "Progress in Measurement of Grain Orientation in Bradford Sand," *Producers Monthly,* 15, No. 8, (1951), pp. 24-26.

11. _____ , "A Further Test of Dimensional Orientation of Quartz Grains in Bradford Sand," *American Journal Science,* 251, (March, 1953), pp. 192-214.

12. Handin, J. W. and D. Griggs, "Deformation of Yule Marble—Predicted Fabric Changes," *Bulletin Geological Society America,* 62, (August, 1951), pp. 863-886.

13. Jizba, Z. V., W. C. Campbell and T. W. Todd, "Core Resistivity Profiles and the Bearing on Dipmeter Survey Interpretation," *AAPG Bulletin,* 48, No. 11, (November, 1964), pp. 1804-1809.

14. Johnson, W. E. and R. V. Hughes, "Directional Permeability Measurements and Their Significance," *Producers Monthly,* 13, No. 1, (1948), pp. 17-25.

15. Kahn, J. S., "Anisotropic Sedimentary Parameters," Transactions, *New York Academy of Sciences,* 21, (1959), pp. 376-386.

16. Krumbein, W. C., "Preferred Orientation in Pebbles in Sedimentary Deposits," *The Journal of Geology,* 47, No. 7, (1939), pp. 673-706.

17. Martinez, J. D., "Photometer Method for Studying Quartz Grain Orientation," *AAPG Bulletin,* 42, No. 3, (1958), pp. 588-608.

18. Mast, R. F. and P. E. Potter, "Sedimentary Structures, Sand Shape Fabrics, and Permeability," *The Journal of Geology,* 71, No. 5, (1963), pp. 548-565.

19. Moran, J. H., M. A. Coufleau, G. K. Miller and J. P. Timmons, "Automatic Computation of Dipmeter Logs Digitally Recorded on Magnetic Tapes," *Transactions AIME,* 225, No. 1, (1962), pp. 771-782.

20. Nanz, R. H., "Grain Orientation in Beach Sands: A Possible Means for Predicting Reservoir Trend," *Journal Sedimentary Petrology,* 25, No. 2, (1955), pp. 127-157.

21. Off, T., "Rhythmic Linear Sand Bodies Caused by Tidal Currents," *AAPG Bulletin,* 47, No. 2, (1963), pp. 324-341.

22. Potter, P. E. and F. J. Pettijohn, *Paleocurrents and Basin Analysis,* New York: Academic Press, Inc., 1963.

23._____and R. F. Mast, "Sedimentary Structures, Sand Shape Fabrics, and Permeability," *The Journal of Geology,* 71, No. 4, (1963), pp. 441-471.

24. Rod, E., "Strike-Slip Faults of Northern Venezuela," *AAPG Bulletin,* 40, No. 3, (1956), pp. 457-476.

25. _____, "Application of Principles of Wrench-Fault Tectonics of Moody and Hill to Northern South America," *Bulletin Geological Society America,* 69, (July, 1958), pp. 933-936.

26. Rusnak, G. A., "Orientation of Sand Grains under Conditions of Unidirectional Fluid Flow," *The Journal of Geology,* 65, No. 4, (1957), pp. 384-409.

27. Schwarzacher, W., "Grain Orientation in Sands and Sandstones," *Journal Sedimentary Petrology,* 21, No. 3, (1951), pp. 162-172.

28. Sriramadas, A., "Appositional Fabric Study of the Coastal Sedimentaries, East Godavari District, Andhra, India," *Journal Sedimentary Petrology,* 27, No. 4, (1957), pp. 447-452.

29. Walton, E. K., "Sedimentary Features of Flysch and Graywackes" in "Developments in Sedimentology," Paper No. 7, Amsterdam: Elsevier Publishing Company, 1965, p. 274.

30. Wyllie, M. R. J. and M. B. Spangler, "Application of Electrical Resistivity Measurements to Problem of Fluid Flow in Porous Media," *AAPG Bulletin,* 36, No. 2, (1952), p. 394.

Paleo-Facies
Logging and Mapping

This type of logging and mapping is of special significance in carbonate-evaporite sediments for lithofacies studies associated with sedimentary environments favorable for reef build-up. In exploring for reefs and for favorable reef building environment, it is often necessary to establish in which facies a well is situated: whether in fore reef, back reef, reef proper, lagoon or basin.

In carbonate-evaporite sequences, oil and gas often accumulate at the up-dip pinch-out of a porous and permeable carbonate formation against a laterally time equivalent tight anhydrite; the pinch-out is due to lateral facies change.

Rock facies are mineral composition changes. Since each mineral has characteristic physical properties such as density, acoustic wave velocity of propagation and neutron response, combining three well logs (density, sonic and neutron) allows the computation of a rock's mineralogic composition and of its porosity—provided no more than three minerals are involved.

The three minerals may include an economically valuable one such as sulfur or potash. Hence, this technique applies to evaluating certain ore deposits.

Lithology of Reefs

Biogenic sedimentary facies that favor reef developments are restricted depositional features schematically represented in Figure

153

6-1. Behind the reef and close to shoreline, evaporites and lagoonal sediments are deposited in a restricted region that generally contributes sediments for approximately 10 miles wide (±) on the basin's shelf side. These sediments are comprised of interbedded green shales and light gray to tan, finely crystalline, anhydritic limestones and dolomites. They grade landward into continental red shales and sands. Toward the reef proper, a dense dolomitic barrier provides a seal. Closer to the reef proper, slabs of reef material (patch reefs) interfinger with back reef carbonates and green shales.

The reef proper is normally devoid of anhydrite, sand and shale. At the front of the reef (fore reef), toward the basin, detritus is deposited by tidal and offshore currents, which mixed and transported the eroded back reef, reef and basin sediments onto the reef's hinge line. Offshore currents may create a depression or trough parallel to shoreline into which debris from wave action may settle and form a reef bank susceptible to becoming a reservoir.

The above brief description of reef reservoirs' genetic process indicates that subsurface exploration for such traps is essentially that of mapping lateral facies or mineral composition changes and outlining their closed configuration. This may be done by combining logs that respond to lithological properties of the rock matrix. Table 6-1 gives the response of various logs to particular lithologies.

Known oil productive reef trends are numerous. In the Permian Basin, to mention just a few, they are in the Canyon, Wolfcamp, Abo and Clearfork formations. In Alberta, Canada, the Leduc reefs are notorious for their prolific oil production.

Lithology of Facies Stratigraphic Traps

Typical facies stratigraphic traps occur in the Williston Basin of North Dakota, Montana and in the Canadian province of Saskatchewan. In this basin, oil pools occur in the Mississippian carbonates as lines of fringing anhydrite sheets along which the

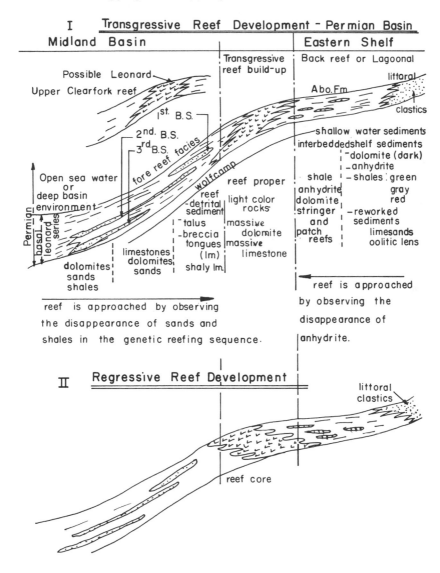

Figure 6-1. General sedimentary and lithologic conditions associated with a transgressive or a regressive reef development.

Table 6-1
Mineral and Rock Responses of Various Well Logs

Mineral or Rock	Symbol as Fraction of Bulk Volume	Sonic Δt: μsec/ft	Sonic V: ft/sec	Density (D) (electron density)	Neutron (H) (indicated porosity as hydrogen index) (GNT tool)	Gamma Ray (as shale contamination)
Limestone	L	45	22,500	2.71	0	0
Dolomite	D	42	24,000	2.87	0	0
Sand (quartz)	Sd	54	18,500	2.65	0 to -4	0
Shales:						
Under-compacted	Sh	>110		1.80	function	function
Normal		100		2.00	of type	of gamma
Compacted		< 90		2.20	shale	deflection
Anhydrite	A	50	20,000	2.98	0	0
Gypsum	G	52.5	19,100	2.35	49	0
Rock salt (halite)	Rs	59-67	15-17,000	2.03	0	0
Sulfur	Su	122	8,250	2.03	?	0
Sylvite	Sy	74	13,500	1.86	0	high disturb-
Polyhalite	Po	57.5	17,400	2.79	15	ance possible
Kainite	Ka	—	—	2.12	45	because of
Carnalite	Ca	78	13,800	1.57	65	radioactive
Langbeinite	La	52	19,200	2.82	0	potassium K^{40}
Water:						
Fresh	W	200	5,000	1.00	100	
100,000 ppmNaCl		189	5,300	1.07	120 approx.	
200,000 ppmNaCl		176	5,700	1.14	140 approx.	

facies changes from porous dolomite to dense anhydrite in less than a mile (Figure 6-2). Three such anhydrite sheets are along the beveled edge of the Mississippian formation, resulting from truncation by a pre-Mezozoic unconformity that cuts the Paleozoic column and which Triassic red beds (Spearfish formation) and other younger formations overlie.

Most reservoir rocks develop in the center section of the Mississippian carbonates (the Mission Canyon formation), which is more than 500 feet thick at various localities. The Mission Canyon is comprised of marine limestones deposited in a westerly regressive environment. As the sea regressed, evaporite sheets

Figure 6-2,a. Lithologic trap in the Madison formation, Williston Basin, North Dakota. (Courtesy of Illing et al. and Elsevier Publishing Company, Ltd.)

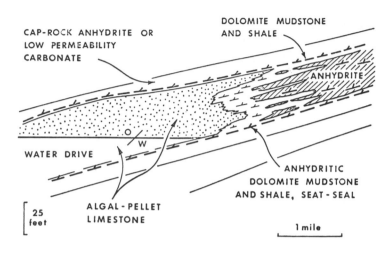

Figure 6-2,b. Lithologic trap in the Madison formation, Williston Basin, North Dakota. (Courtesy of Illing et al. and Elsevier Publishing Company, Ltd.)

encroached toward the basin's center and overlapped the carbonates, thus forming a lateral seal to up-dip fluid migration and a vertical trap for hydrocarbons because of permeability's disappearing laterally and vertically—i.e., from porous dolomite to tight anhydrite. Attitude of the facies contrast changes according to the different lithologies, but this contrast is the important element in forming stratigraphic traps in the Williston Basin.

In contradistinction to stratigraphic traps in sand-shale sediment sequences, geologic time lines do not confine stratigraphic traps because boundaries between facies cut across time lines. Facies changes are time regressive. Hence, these stratigraphic traps may not be defined by mapping time parallel marker beds or genetic time strata interval, but rather by mapping rock types (i.e., areas of changes among rock compositions). To define such a trap's sealing edge, one needs to map the contact from carbonate to anhydrite; to define the porosity edge, one needs to map the contact from limestone to dolomite.

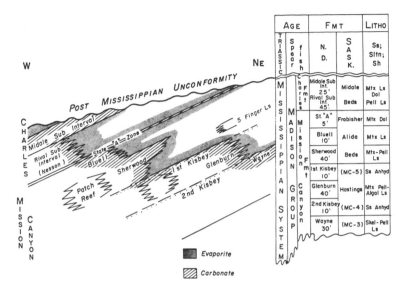

Figure 6-2,c. Lithologic trap in the Madison formation, Williston Basin, North Dakota. (Courtesy of W. W. Wakefield and *The Oil and Gas Journal.*)

Major oil fields so generated in the Williston Basin include, for example, Weyburn, Midale, Steelman, Alida, Parkman and Wiley. Similar principles have helped define stratigraphic traps in the San Andres formation of the Permian Basin, in which major oil reserves such as the Levelland-Slaughter and Chaveroo fields have also been found.

Along the Mexia-Talco fault zone, Smackover production (oil, gas, distillate, hydrogen sulfide) is also associated with a facies change from tight limestone to porous dolomite.

Three Porosity Logs

As Table 6-1 shows, responses of the three porosity logs (sonic, density and neutron) are individually different for each mineral entering rock composition. It is apparent, therefore, that combining three porosity tools with a gamma ray curve may solve the problem of having three minerals plus shale contamination, since four log curves (i.e., four readings at any one stratigraphic level) help solve four simultaneous equations at that level. As the sum of the fractions of each mineral plus porosity must equal "1" an additional equation is provided.

Porosity, three minerals and shale contamination, may be calculated when three porosity logs and gamma ray-neutron logs exist. Quantitatively evaluating the above five unknowns requires pre-identifying the minerals by core or sample mineralogic examination. The following set of simultaneous equations may be written:

Sonic Log

$$\Delta t_{log} = \phi \ [S_w \Delta t_w + S_o \Delta t_o + S_g \Delta t_g] + \Sigma(MF) \cdot \Delta t_m \quad (6\text{-}1)$$

Density Log

$$D_{log} = \phi \ [S_w D_w + S_o D_o + S_g D_g] + \Sigma(MF) \cdot D_m \quad (6\text{-}2)$$

Neutron Log

$$N_{log} = a - b \cdot \log H \quad (6\text{-}3)$$

wherein:

Hydrogen Index

$$H = \phi \ [S_w H_w + S_o H_o + S_g H_g] + 0.49 \ G + S_h H_{sh} \quad (6\text{-}4)$$

where:

$$S_h = \frac{GR \ log}{GR_{sh}} \quad (6\text{-}5)$$

and:

$$H_{sh} = \frac{2.65 - D_{sh}}{2.65} \quad (6\text{-}6)$$

Saturation

$$1.0 = S_w + S_o + S_g \quad (6\text{-}7)$$

Bulk Volume

$$1.0 = \phi + \Sigma(MF) \quad (6\text{-}8)$$

Identity
$$S_d = 0.0\,\phi + 0.0\,A + 0.0\,D + 0.0\,G + 1.0\,x\,S_d \qquad (6\text{-}9)$$

to introduce silica (S_d) as needed in eliminating negative values for various mineral components. If silica is not a likely accessory mineral, properties of another more likely mineral should be tried.

Symbols and subscripts in the above equations have these meanings:

Symbols
> Δt : sonic transit time : μ sec / ft
> D : density : grm / cm^3
> N : neutron deflections : units arbitrary
> H : hydrogen index specific to the neutron-gamma log run
> MF: mineral fraction in total rock bulk volume
> Sh : shale fraction in total rock bulk volume
> ϕ : porosity fraction in total rock bulk volume
> G : gypsum fraction in total rock bulk volume
> S_d : sand fraction in total rock bulk volume

Subscripts
> log : property recorded on the log
> w : property of formation water
> o : property of reservoir oil
> g : property of reservoir gas
> m : property of specific mineral
> sh : property of shale

Log Characteristics
> a : neutron deflection for a 1.0 hydrogen index
> b : slope of neutron-porosity calibration curve specific to the neutron log run; "b" varies with hole size, neutron-detector spacing, neutron source, whether the hole is cased, etcetera. (The semi-logarithmic relation in equation 3 may not be valid at very low porosities.)

For a complex problem, using computers is the best way to solve the above nine simultaneous equations at various stratigraphic levels. However, the ordinary explorationist seldom knows enough about computers and solving simultaneous equations for a discussion of this subject to be beneficial. On the other hand, it is often necessary for him to check results of computer calculations or to make certain calculations himself. Then, the problem must be simplified, and simple charts are available for calculating mineral facies and identifying matrix minerals.

When only two minerals are in the matrix and the pore fluid density may be considered to be near unity ($S_g = 0$, $D_o = D_w$ and $H_o = H_w$), cross plots of two porosity tools should then identify the two minerals and determine the porosity value. Such charts are in Figures 6-3, 6-4 and 6-5. The appropriate chart is selected according to which porosity tool combination is used. A cross plot of log readings at a given level defines a point on the graph; and, if the two minerals are known, the respective amount of each is obtained by interpolation and the porosity is read.

Using three porosity tools and knowing that the three minerals most likely to be present are dolomite, gypsum and anhydrite and the mean pore fluid density is unity, then the chart of Figure 6-6 can solve equations at each stratigraphic level. The chart is first entered at the upper left with Δt, joining ρ and then ϕ_n as shown. This last line determines G (gypsum) and ϕ (porosity). Then, the chart at the bottom is entered from the left with ϕ_n, ρ, Δt and G (previously determined) and also from the right with ϕ (previously determined). Thus, the dolomite and anhydrite fractions are determined. This chart does not permit adjustments when negative values are obtained. However, negative values indicate that assuming the presence of a given mineral is wrong and a likely one of higher or lower density should be tried.

Special Cases of Computer Application
to Lithology Logging

Reefing Environment

A common lithology with respect to reef exploration is a rock containing the following minerals in various proportions in addition to porosity (ϕ):

D : dolomite
A : anhydrite
G : gypsum
Sd: silica (as in Table 6-1)

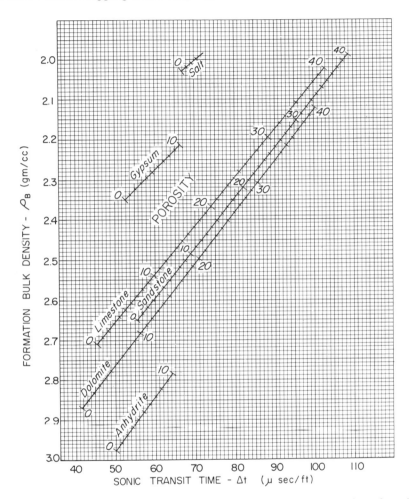

Figure 6-3. Lithology and porosity determination from crossplot of sonic transit time versus formation bulk density. (Courtesy of Schlumberger Well Services.)

These equations will yield acceptable rock composition:

Δt = 185.0 ϕ + 42.0 D + 50.0 A + 53.0 G + 55.0 Sd (sonic)

ρb = 1.10 ϕ + 2.84 D + 2.98 A + 2.35 G + 2.65 Sd (density)

ϕ_n = 1.0 ϕ + 0.02 D + 0.0 A + 0.49 G –0.04 Sd (neutron)

1.00 = 1.0 ϕ + 1.0 D + 1.0 A + 1.0 G + 1.0 Sd (unity)

Sd = 0.0 ϕ + 0.0 D + 0.0 A + 0.0 G + 1.0 Sd (identity)

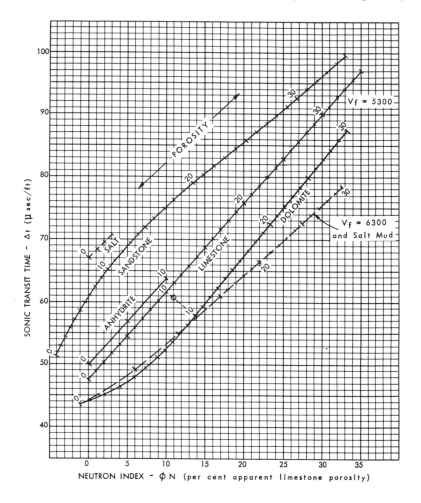

Figure 6-4. Lithology and porosity determination from crossplot of sonic transit time versus neutron porosity index. (Courtesy of Schlumberger Well Services.)

where: Δt, ρ_b, and ϕ_n = input values from sonic, density and neutron logs, respectively

ϕ = porosity

D, A, G and Sd = computed fractions of bulk volume attributed to dolomite, anhydrite, gypsum and silica, respectively.

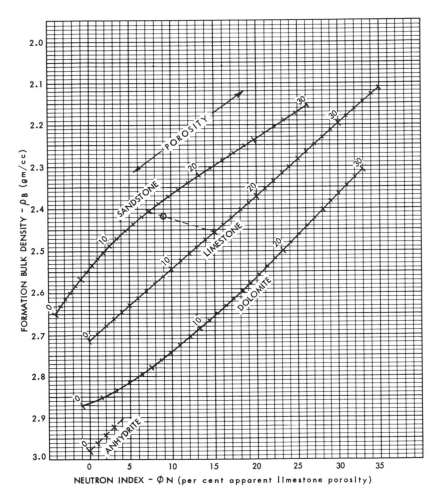

Figure 6-5. Lithology and porosity determination from crossplot of formation bulk density versus neutron porosity index. (Courtesy of Schlumberger Well Services.)

Five unknowns and only four basic equations exist. To resolve the difficulty, the "identity" equation has been added and is used only when a negative answer for one of the minerals is obtained. Then, "silica" is added systematically to the matrix mixture, one percent or less at each step, until negative answers are no longer obtained. While the answer is not necessarily correct or the best

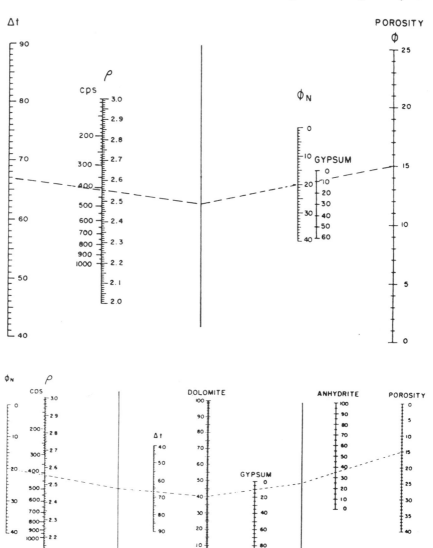

Figure 6-6. Nomograph for determining lithology and porosity when the rock's three minerals are dolomite-gypsum and anhydrite. (Courtesy of Schlumberger Well Services.)

answer, acceptable results have been obtained on the basis that porosity changes very little by adding silica.

Typical results of such computer calculations are in Figure 6-7, showing that statistical means may arrive at the most probable composition when testing eight possible mineral components plus porosity. Solving the four simultaneous equations yields 70 possible answers at each foot in depth (Dawson-Grove and Palmer[4]). At every level, the eight unknowns are taken four at a time. Certain constraints are placed in the solution, and the surviving solutions are examined for the minerals which occur most frequently. Gamma ray curve also monitors the surviving solutions and picks the best ones. The eight components tested in this program are *porosity, limestone, shale, dolomite, anhydrite, salt, gypsum and iron minerals (limonite, siderite, pyrite)*.

Dolomitic Limestone of Low Porosities

A typical reservoir rock of this type is the deep Ellenburger of the Delaware Basin, West Texas. Porosity averages only slightly over 2 percent in good reservoirs, and extreme accuracy is required in formation evaluation. While the Ellenburger is primarily a dolomite, small fractions of other minerals are in the matrix and have been identified primarily as anhydrite and calcite. The bore fluid and saturation degree are also important especially because the gas density greatly differs from that of water. At extreme pressures, gas specific gravity is 0.15 and water specific gravity is 1.15 because of high salinity and pressure. Simultaneous equations for solving the Ellenburger problem are the following, according to Neustaedter[14]:

$$\Delta t = 156.0\,\phi_w + 156.0\,\phi_g + 42.0\,D + 50.0\,A + 47.0\,L \text{ (sonic)}$$
$$\rho_b = 1.15\,\phi_w + 0.15\,\phi_g + 2.87\,D + 3.0\,A + 2.71\,L \text{ (density)}$$
$$\phi_n = 1.00\,\phi_w + 0.5\,\phi_g + 0.02\,D + 0.0\,A + 0.0\,L \text{ (neutron)}$$
$$1.00 = 1.00\,\phi_w + 1.0\,\phi_g + 1.0\,D + 1.0\,A + 1.0\,L \text{ (unity)}$$
$$L = 0.0\,\phi_w + 0.0\,\phi_g + 0.0\,D + 0.0\,A + 1.0\,L \text{ (identity)}$$

where: $\Delta t, \rho_b, \phi_n =$ input from sonic, density and neutron logs, respectively

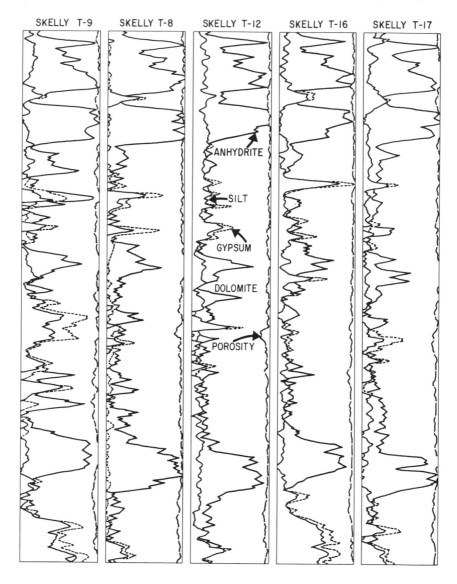

Figure 6-7. West-east correlation of computed lithologic logs that illustrates lateral variation in rock mineral composition and reservoir thickness. (Courtesy of J. A. Burke, et al., and the Society of Petroleum Engineers of AIME.)

> D, A, L   = percentages of dolomite, anhydrite and lime, respectively
>
> ϕ_w = water-filled porosity
>
> ϕ_g = gas-filled porosity.

Rocks with Shale Contamination

In this eventuality, the gamma ray curve must evaluate the shale contamination. The technique assumes that the contaminating shale has the same properties (porosity and radioactivity) as the well-defined shales above and below the particular formation. In sands this may be an untenable assumption, as contaminating shale fractions (p or q) are often more radioactive than adjacent shales. In carbonates, however, the above assumption is more nearly correct.

The above nine equations are then solved simultaneously at every foot interval from log readings and are entered in the computer from data cards or digitized logs.

Mosa[12] has designed the computer program of Appendix 2 to solve this particular problem.

Lithologic Logging with Only Two Porosity Curves and Gamma Rays

Porosity curves may be combined accordingly: sonic-density; sonic-neutron; neutron-density. Theoretically, it is then possible to solve only for two minerals—shale contamination and porosity.

However, Mosa[12] has designed a computer technique that uses two identity equations and programs trial and error solutions to try out the likely amount of contaminating minerals.

Assume, for example, that sonic and neutron logs and the gamma ray log are available. The following equations may be written:

$$\Delta t = 186.0\,\phi + 42.0\,D + 50.0\,A + 105\,Sh + 53.G + 55.0\,Sd$$
$$\phi_n = 1.0\,\phi + 0.02\,D + 0.0\,A + 0.34\,Sh + 0.49\,G - 0.04\,Sd$$
$$1.0 = 1.0\,\phi + 1.0\,D + 1.0\,A + 1.0\,Sh + 1.0\,G + 1.0\,Sd$$

GR = 0.0 ϕ + 0.0 D + 0.0 A + 143.0 Sh + 0.0 G + 0.0 Sd
G = 0.0 ϕ + 0.0 D + 0.0 A + 0.0 Sh + 1.0 G + 0.0 Sd
Sd = 0.0 ϕ + 0.0 D + 0.0 A + 0.0 Sh + 0.0 G + 1.0 Sd

"Gr is the gamma ray deflection on the log at a certain level. For the contaminating shale, this deflection is 143.0 API units. The last two equations are identity equations, which are satisfied by introducing the required amount of contamination in gypsum (G) and silica (Sd). Appendix 2 contains computer details of this technique.

Evaluating Native Sulfur Deposits

Native sulfur occurs with two different kinds of source rock.

Anhydrite. In the cap rock of salt domes, it occurs with anhydrite, which was the mineral reduced to native sulfur by sulfate-reducing bacteria. Bacteria use hydrocarbons leaking around the salt dome as their energy source, their metabolic process then yields secondary calcite besides free sulfur. The rock combination is anhydrite (A), calcite (L), sulfur (Su), porosity (ϕ). Then, equations to be solved simultaneously are

Δt = 122 Su + 47.5 L + 50.0 A + 189 ϕ
ρ_b = 2.03 Su + 2.71 L + 2.98 A + 1.0 ϕ
ϕ_N = 0.0 Su + 0.0 L + 0.0 A + 1.0 ϕ
1.0 = 1.0 Su + 1.0 L + 1.0 A + 1.0 ϕ

Gypsum. In evaporite basins—such as, the Delaware Basin of West Texas or the Sicilian Sulfur deposits—the association of free sulfur is primarily with gypsum, although anhydrite is not necessarily absent. Native sulfur originates from gypsum again by sulfate-reducing bacteria and generation of secondary calcite. The rock combination is gypsum (G), calcite (L), sulfur (Su), porosity (ϕ).

Then, equations to be solved simultaneously are

Δt = 122 Su + 47.5 L + 52.5 G + 189 ϕ
ρ_b = 2.03 Su + 2.71 L + 2.35 G + 1.0 ϕ
ϕ_n = 0.0 Su + 0.0 L + 0.49 G + 1.0 ϕ
1.0 = 1.0 Su + 1.0 L + 1.0 G + 1.0 ϕ

Figure 6-8 illustrates results of lithologic interpretation from a sulfur deposit (according to Tixier and Alger[21]).

Evaluating Potash Deposits

"Potash" applies to potassium-bearing minerals generally found in evaporite sediments. Because radioactive isotope K^{40} is 0.012 percent of the total potassium fraction, all potash deposits are radioactive. Hence, without other disturbing radioactive minerals, it is possible to evaluate the K_2O which may be at all levels in a well: When the gamma ray curve is properly calibrated and when hole diameter, logging speed and circuit time constant are allowed for, it is possible to determine a true value of the radioactivity and to use charts (i.e., Figure 6-9) to evaluate K_2O at all depth levels.

In practice, potash deposits mix various potassium-bearing minerals (as given in Table 6-1) with other nonradioactive evaporites and nonsoluble minerals. The most common mixture encountered may include *potash minerals*, sylvite (Sy) and carnalite (Ca); *sodium salt*, halite (Rs); *insolubles*, Anhydrite, gypsum, calcite and dolomite (I).

Because of the mixture's complexity, the physical properties of insoluble minerals must be grouped into single values, as Tixier and Alger suggest[21], like the following:

$$\Delta t_i = 120.0 \text{ ft/sec}$$
$$\rho_{bi} = 2.6 \text{ gm/cm}^3$$
$$\phi_{ni} = 0.3$$

and tried out as the accessory mineral in these equations with the physical properties of Table 6-1:

$$\Delta t = 185.0\,\phi + 74.0 \text{ Sy} + 78.0 \text{ Ca} + 67.0 \text{ Rs} + 120.0 \text{ I}$$
$$\rho_b = 1.1\,\phi + 1.86 \text{ Sy} + 1.57 \text{ Ca} + 2.03 \text{ Rs} + 2.6 \text{ I}$$
$$\phi_N = 1.0\,\phi + 0.0 \text{ Sy} + 0.65 \text{ Ca} + 0.0 \text{ Rs} + 0.3 \text{ I}$$
$$1.0 = 1.0\,\phi + 1.0 \text{ Sy} + 1.0 \text{ Ca} + 1.0 \text{ Rs} + 1.0 \text{ I}$$
$$I = 0.0\,\phi + 0.0 \text{ Sy} + 0.0 \text{ Ca} + 0.0 \text{ Rs} + 1.0 \text{ I}$$

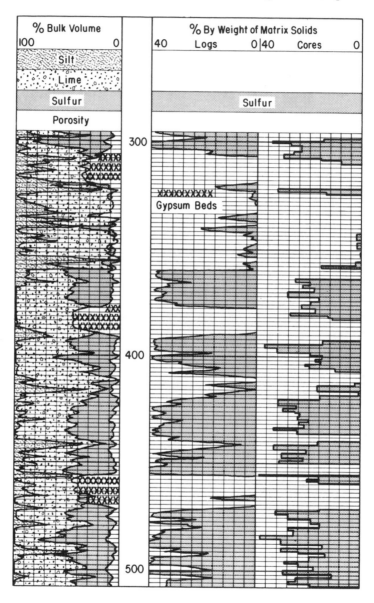

Figure 6-8. Lithology and sulfur content computed from three-porosity log combination in a sulfur test hole near Fort Stockton, Pecos County, Texas. (Courtesy of M. P. Tixier, R. P. Alger and SPWLA.)

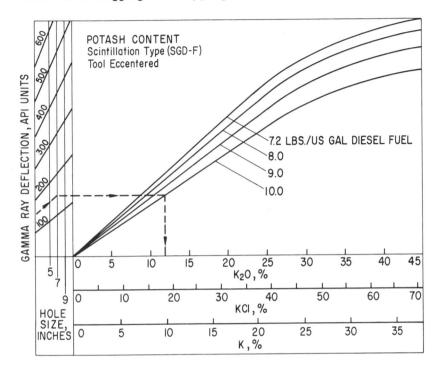

Figure 6-9. Empirical chart relating gamma ray deflection to apparent potassium content. (Courtesy of Schlumberger Well Services.)

where:

Sy = sylvite fraction

Ca = carnalite fraction

Rs = rock salt or halite fraction

I = insoluble fraction.

Computer calculations would then be carried out as usual.

Other Paleo-facies Mapping Techniques

McCrossan[11] has observed marked and significant changes in the resistivity of Ireton shale near the Leduc reef fields in Alberta, Canada. The shale resistivity is highest near the reefs and lowest to normal 8 to 10 miles from them. Figure 6-10 reproduces McCrossan's facies study of the inter-reef calcareous Ireton shale

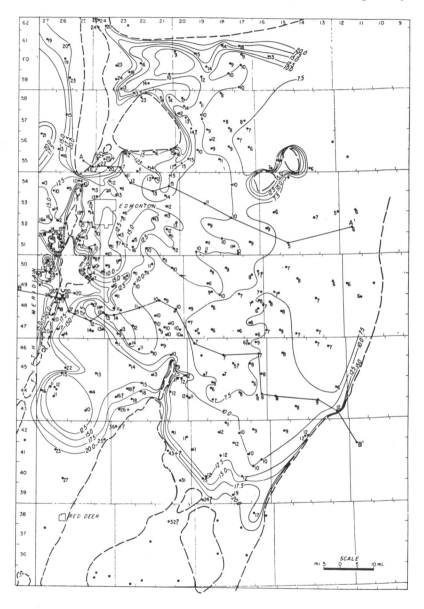

Figure 6-10. Resistivity map of the middle and lower Ireton shale, Alberta, Canada, as an index of the proximity to Leduc reef build-up. (Courtesy of R. G. McCrossan and AAPG.)

based on its resistivity from the short normal curve of the area's well logs. Resistivities vary from 20 Ωm near the reefs to 7.5 Ωm in the depositional basin's central part. This implies that the Ireton shale's calcisiltite was reef-derived as a fine lime powder by destructive seawaves while the reefs were building upward. This fine clastic carbonate powder spread throughout the Ireton Sea decreasingly as the distance from the reef pinacles increased.

Of course, difficulties confront the significance of such studies, as the shale resistivity could vary not only with carbonate content but with variations in other mineralization (i.e., clay content, organic content, gypsum, anhydrite, sand, water salinity and differential compaction). The idea is, nevertheless, very interesting and may perhaps be improved by lithologic logging in wells with three porosity curves. It would then be possible to calculate the Ireton shale's mineralogic composition—especially, significant minerals such as anhydrite, dolomite, calcite and quartz. Areally mapping the composition would undoubtedly be more significant than resistivity mapping. A combined parameter, like one derived from a triangular plot and on which the distance from the shale composition's representative point to one representing the reef's mineral composition, could be called a "reef proximity index."

By analogy, in sand-shale sequence, stratigraphic traps may be mapped from a sand proximity index derived from lithologic logs interpreted by quartz, clays and feldspars. Resistivity mapping of a strata's genetic interval may also be an exploration tool for mapping a relative amount of coarse sedimentary grains.

Example of Application

The three-porosity log in Figure 6-11 is from a San Andres well, Crane County, Texas, and reproduced from Savre.[18] Calibration of the curves did not appear in the article and were inserted on the logs according to the area's usual logging practice. It is desired to compute the lithologic composition of the rock at levels 1 and 2. While these levels are somewhat shaly, as the gamma ray curve shows, it is not possible to evaluate the degree of shale contamination, as the log does not extend into adjacent shales.

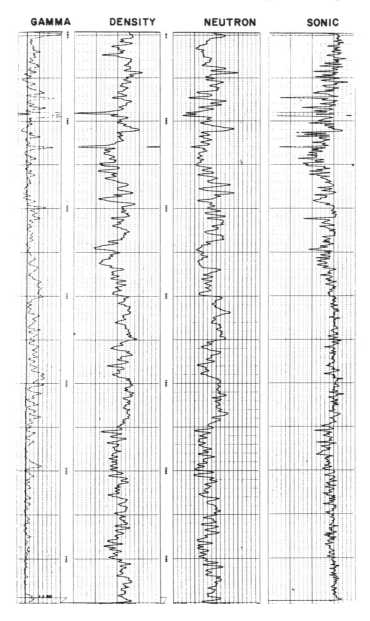

Figure 6-11. Example of three porosity logs (gamma ray, density, neutron and sonic in a San Andres well, Crane County, Texas. (Courtesy of W. C. Savre and the SPE of AIME.)

Calculations will then be performed according to Figure 6-6, since the San Andres Reservoir rock is mostly anhydrite, gypsum and dolomite. The following are the results:

Levels	Readings			Computed Values			
	D	ϕ_N	Δt	ϕ	D	G	A
1	2.7	10	64	13.0	35	44	8
2	2.7	18	61	10.5	0	15	75

References

1. Burke, J. A. et al., "Computer Processing of Log Data Enables Better Production in Chaveroo Field," *Transactions of the Society of Petroleum Engineers,* (July, 1967), pp. 889-895.

2. Burnside, R. J., "Geology of Part of the Horseshoe Atoll in Borden and Howard Counties, Texas," USGS Prof. Paper 315-B, 1959.

3. Chester, R., "Geochemical Criteria for Differentiating Reef from Non-Reef Facies in Carbonate Rocks," *AAPG Bulletin,* 49, No. 3, (March, 1965), pp. 258-276.

4. Dawson-Grove, G. E. and K. R. Palmer, A Practical Approach to Analysis of Logs by Computer," unpublished paper, Home Oil Company Ltd. (Calgary, Alberta).

5. Edwards, J. M. et al., "Nuclear Log Evaluation of Potash Deposits," unpublished paper, McCullough Tool Company.

6. Gratton, P. J. F. and W. J. LeMay, "New Mexico Search for San Andres Permian Oil Most Active in State," *The Oil and Gas Journal,* 65, No. 12, (March 20, 1967), pp. 207-212.

7. Illing, L. V. et al., "Reservoir Rocks and Stratigraphic Traps in Non-reef Carbonates," *Proceedings Seventh World Petrolem Congress,* Mexico City, PDOC 4, 1967, pp. 49-75.

8. Ingerson, E., "Problems of the Geochemistry of Sedimentary Carbonate Rocks," *Geochimica et Cosmochimica Acta,* 26, (1962), pp. 815-847.

9. Klovan, J. E., "Facies Analysis Aids Reef Exploration and Development," *World Oil,* 160, No. 6, (May, 1965), pp. 141-150.

10. Lebreton, F. et al., "Log-Combining Refines Porosity Measurements," *Petroleum Engineer,* 35, No. 13, (December, 1963), pp. 88-113.

11. McCrossan, R. G., "Resistivity Mapping and Study of Upper Devonian Inter-Reef Calcareous Shales of Central Alberta, Canada," *AAPG Bulletin,* 45, No. 4, (April, 1961), pp. 441-470.

12. Mosa, M. M., "Computer Analysis of Lithology Logging," unpublished Master's thesis, The University of Texas, June, 1968.

13. Myers, D. A. et al., "Geology of the Late Paleozoic Horseshoe Atoll in West Texas," No. 5607, Texas Bureau of Economic Geology, April, 1956.

14. Neustaedter, R. H., "Log Evaluation of Deep Ellenburger Gas Zones," SPE Paper 2071, March, 1968.

15. Nottingham, M. W., "Abo Reef Build-up Provides Five Stratigraphic Trap Zones," *World Oil,* 162, No. 6, (May, 1966), pp. 107-110.

16. Raymer, L. L. and W. P. Biggs, "Matrix Characteristics Defined by Porosity Computations," Paper X, SPWLA Symposium, Oklahoma City, May, 1963.

17. Runyan, J. W., "First New-Mexican Reef Detritus Oil Pools Found Down Dip From Abo Trend," *World Oil,* 160, No. 4, (March, 1965) pp. 99-106.

18. Savre, W. C., "Determination of a More Accurate Porosity and Mineral Composition in Complex Lithologies with the Use of the Sonic, Neutron, Density Surveys," *AIME Journal of Petroleum Technology,* September, 1963, pp. 945-959.

19. _____ and J. A. Burke, "Determination of True Porosity and Mineral composition in Complex Lithologies with the Use of the Sonic, Neutron, and Density Surveys," Paper XI, SPWLA Symposium, Oklahoma City, May, 1963.

20. Stafford, P. T., "Geology of Part of the Horseshoe Atoll in Scurry and Kent Counties, Texas," USGS Prof. Paper 315-A, 1959.

21. Tixier, M. P. and R. P. Alger, "Log Evaluation of Non-metallic Mineral Deposits," Paper R, SPWLA Symposium, Denver, June, 1967.

22. Wakefield, W. W., "Much Good Oil-hunting Land Left in North Dakota's Mississippian," *The Oil and Gas Journal,* 63, No. 22, (May 31, 1965), pp. 138-146.

23. Wright, W. F., "Potential West Texas Giant (Capitan Reef) Ignored," *The Oil and Gas Journal,* 65, No. 25, (June 19, 1967), pp. 176-178.

Fracture Intensity
Logging and Mapping

Many reservoir rocks—such as chalk, tight sands, limestones, dolomites, shales and schists—are so devoid of permeability that no production can be achieved unless tectonic deformation processes have naturally fractured them—even though some may have high porosity and may contain large quantities of oil per acre-foot. Economic production still is not attained unless artificial fracturing is resorted to.

These are examples of such reservoir rocks:

          Annona chalk—Pine Island Field, Louisiana
          Selma chalk—Louisiana and Alabama
          Austin chalk—Salt Flat Field, Texas
          Spraberry sand—Texas
          Sanish sand—Antelope Field, North Dakota
          Dakota shale—Florence Field, Colorado
          Basal schist—San Joaquin Basin, California

Natural fracturing generally occurs from tension or shear stresses in a competent or brittle bed, and the frequency or intensity of shattering depends and is associated with the region of maximum deformation. Major tension faults are also generated by such deformations, and displacement and drag along the faulting surface generate additional shattering. The intensity of this shattering decreases as distance from the fault plane increases. The most oil productive regions of a fractured field are shattered the most. Thus, a way to measure the "fracturing intensity" and to map this parameter in guiding an exploration program toward regions of expected maximum productivity should be developed.

The fracture intensity parameter may not only be mapped on the reservoir rock but on an associated brittle layer in its vicinity; and it also may be mapped indirectly from fractures' being more frequent and open when the rocks are more tectonically deformed (as the curvature of deformed surfaces may measure). The following describes a computer curvature mapping technique that maps areas of highest fracture intensity.

Fault Recognition

Since rock fracturing is also associated with faults, such faults should be recognized in the sedimentary sequence. In petroliferous provinces, faulting has to be recognized mostly in shale sequences, as they constitute more than 75 percent of sedimentary rocks. Hence, detecting and recognizing faulting rest on recognizing correlatable shale sequences.

Well logs have already proved invaluable to the subsurface stratigrapher in supplying him with a duplicatable and objective record of some of the significant physical properties of rocks (i.e., the sample description of color, grain size, cementation and texture), which are mostly devoid of human inconsistencies. Gross petrophysical characteristics have proven consistent over large areas. Frequently, some obvious and recurring patterns of curve "kicks" on the SP curves or on the resistivity and radioactivity curves are called "fingers" (one, two, three, etc.), depending on the number occurring in close association. Often they may be correlated in an obvious and unquestionable manner over extensive parts of a sedimentary basin. Porous sands and limestones exhibit characteristic features, which are readily detected and recognized.

This discussion concentrates more on the less obvious sedimentation patterns—those seemingly immersed in random noise that the bore hole, instrumentation, atmospheric and man-made disturbances cause plus nonreproducible rock inhomogeneities. Such low level variations occur mostly in shale sections, which are also the most prominent sedimentation bodies in a basin. While variations of physical shale properties are often at a low level of magnitude, they can and should be properly amplified as sedimentation markers. This is particularly desirable for the short normal curve (AM=16″). The amplified short normal, or "correlation curve," is a valuable sedimentation tool.

Usually, mathematical techniques of auto-correlation and cross-correlation applied to digitized logs are not needed to recognize fundamental sedimentation patterns within shale bodies. This discussion is concerned with such obvious patterns, although the process may be completely computerized, thus permiting pattern messages believed to represent significant geological cycles to be extracted from highly disturbed or poor quality logs.

Since the significant sediments in oil and gas exploration were deposited on sloping continental shelves in a highly fluid state, associated shale masses in the formation process often reached static disequilibrium; large volumes of it traveled downward far as "turbidity currents," or "mud flows," which moved along the bottom ocean slope in standing deep water. Because of instability or a stimulus that an earthquake, for example, causes, unconsolidated material on a slope may begin to slide down the continental shelf.

Evidence shows that mud flows are relatively common in shale sections. They most likely occur where sediments are accumulating rapidly on the edge of steep submarine slopes bordering the continents. Mud flows can supposedly take place where the angle of slope is only about one degree, but this does not imply that slopes of greater angle are devoid of unconsolidated materials. The frequency of mud flows is expected to be greater where active faults (gravity or contemporaneous) develop during sedimentation, as was the case for sediments in the Gulf of Mexico from Cretaceous time to the present.

When mud flows attain high speed (i.e., up to 60 miles per hour), they are called "turbidity currents," a terminology P. H. Kuenen proposed in 1939. Soft sediments—when suspended in great quantities of water (from unusual agitation by storm waves, earthquakes, etc.)—form a highly dilute mud, which is denser than sea water and which tends to flow down continental slopes by gravity. If the flow develops a sufficiently high translation velocity, turbulence sets in; for example, violent and random velocity fluctuations occur about the main velocity. These fluctuations are irregular in magnitude and direction.

When turbulent flow develops, resistance to flow by friction on sea bottom diminishes; in fact, the current erodes much material from sea bottom and the current load thereby increases. Because of turbulence, the current can transport much more suspended material than laminar flow. While no one has ever seen a turbidity current, the phenomenon is analogous to a snow avalanche down mountain slopes, the velocity of which may attain hundreds of miles per hour; or to "cinder clouds" that almost instantaneously descend the slopes of erupting volcanoes (i.e., Herculaneum and Pompei, Mount Pelée), killing all life over large areas.

Mathematics and physical models have shown that such high travel speeds of a turbidity current are possible. One such current seemingly occurred in 1929, but the phenomenon was not interpreted as a mud suspension flow until 1952 by B. C. Heezen and M. Ewing. This was the Grand Banks earthquake south of Newfoundland. These authors reasoned that massive slope material slumped south onto Telegraph Plateau, causing five transatlantic cables to rupture successively in a few hours. Since the position of each cable and the exact time at which telegraph services were interrupted in each were known, it was possible to calculate the speed of the suspended mud flow at 60 miles per hour. The "turbidite" formation that this event laid out has been mapped partially and appears as a tongue 300 miles long by 200 miles wide, which transported 100 Km^3 of material.

This phenomenon has been dwelt on considerably in order to establish its possibility and the expectancy of its frequent occurrence in tectonically active regions but, more specifically, in association with active depocenters, such as delta regions on the edge of continental shelves. Kuenen has stated that the time interval between turbidity currents may range from 500 to 10,000 years.

True turbidites are abyssal deposits, but many other gravity sediments never attain a true turbidite status. Yet, they are due to sudden displacement of large fine-grained masses, the spreading of which is confined to restricted areas and may be considered as thick chaotic slump material that Flores[6] named "olistostrome." These slump blocks are intercalated within more normal type sediments and may be classified as their own stratigraphic units. Because of their precipitation, these thick slump blocks resemble turbidites—namely, in graded bedding—but over thicker sections (Figures 7-1 and 7-2).

Log Characteristics of Turbidites and Slump Blocks

We are concerned mainly with mud flow sediments within shale masses. Figure 7-1 illustrates a low level negative SP deflection at the base which gradually tapers toward the top. On the resistivity side, very low resistivity appears at the base. The two combined deflections indicate that coarser grain material is at the base; and the grains gradually become much finer toward the top, as the reduced SP and the higher resistivity indicate. These observations coincide with turbidite properties, which are described as "graded bedding" or stratification sequence, in which each stratum displays a gradation in grain size from coarse (below) to fine (above). Grading of grains comes from the almost instantaneous character of deposition from suspension. Thus, the orientation of elongated grains should freeze in their relative positions when suspended.

Turbidites are anisotropic in their physical properties, and preferred dimensional grain orientation imparts sufficient anisotropy in rock strength to influence fracture orientation. However, J. H. Spotts found that the directional features of the Topanga sandstone, California, were 35 degrees counterclockwise from the down current's strike direction. These observations are not conclusive until many more measurements are made. However, the lack of agreement possibly may be real and due to the Coriolis force when the grains were suspended.

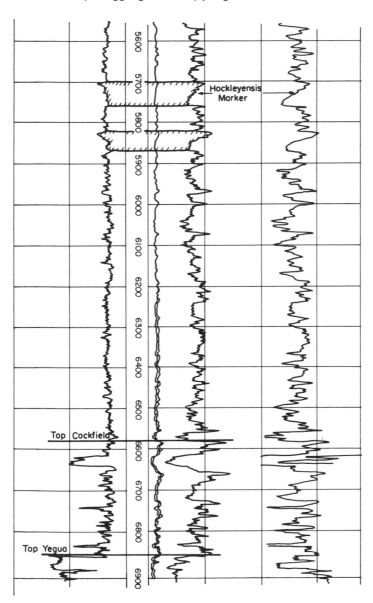

Figure 7-1. Example of widely correlatable slump blocks in the Jackson shale, Liberty County, Texas.

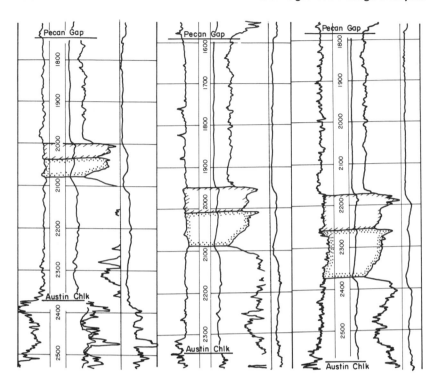

Figure 7-2. Examples of widely correlatable slump blocks in the Taylor shale, Caldwell County, Texas.

True turbidite sedimentary layers are rather thin, one inch to 10 feet up, as in abyssal plains of the deep seas. However, by successive and repeated deposition plus interbedded pelagic sediments, they are believed to form great sediment thicknesses, which occasionally become exposed as outcrops like those in the Maritime Alps, the Marathon Uplift, Texas; and in California. The original departure point as source suspension may be hundreds of miles away, and part of the original mud flow may be deposited closer to the source with characteristics not of a true turbidite. However, the recurring phenomenon may cause great thickness of bedded fine grain sediments of extensively similar characteristics. It is difficult to visualize other sedimentation mechanisms that would have produced such characteristics and which are duplicated so faithfully in well log shale sections (Figures 7-1 and 7-2).

The logs of Figure 7-1, in the 5,700 to 5,800 foot interval (immediately below the Textularia hockleyensis fossil marker) show a resistivity pattern which is reproducible over Liberty County, Texas, and some adjacent counties. Other recurring patterns appear on the same logs. For easy detection, identification, recognition and correlation within shale bodies, it is suggested to color each identified paleo-electric marker between the amplified normal and the SP curve and to use a different color for each slump block.

The logs of Figure 7-2 are from Caldwell County, Texas, near the Luling-Mexia fault zone. They show similar reproducing patterns in the Taylor shale, some of which are hachured for easy identification and correlation.

The changing character of each unit that the envelope of the SP curve and the amplified normal enclose may thereby be studied as to thickness and grading, leading to some conclusion about the shape and extent of a slump block, its probable source area and possibly about the location of earth movements (fault displacements) which trigger the mud flow. The proximity and intensity of the associated diastrophism may be inferred from the individual units' shape and graded character. This appears possible along the Luling-Mexia fault zone, and the proximity to zones of intense fracturing may be evaluated accordingly. This approach appears most significant to find economically oil producible reservoirs in the Austin chalk and in the Smackover dolomite. These formations do not become economic reservoirs unless intensely fractured.

Detecting Subsurface Faulting

Moore[12] lists criteria for detecting subsurface faulting. Only these are strictly observable from logs and mapping:

1. Dip of a marker bed abruptly changes on a cross section based on logs or on a contour map of a marker bed.

2. Omission of marker beds generally spells a normal fault regardless of whether it is (a) a gravity type with the downthrown side toward the basin (Gulf Coast growth faults) or (b) a cantilever type with the upthrown side toward the basin (Luling-Mexia faults).

3. Repetition of marker beds characterizes thrust faulting (South Oklahoma).

4. Thickening of marker beds on the fault's downthrown side characterizes growth faults.

By adequately correlating and identifying shale sections from logs as indicated above, the bed thickness (omitted or repeated) can be evaluated exactly—thus, to evaluate the vertical throw distance.

Fracture Intensity Mapping

The earth pulsates an average of 10 to 15 inches, four times a day under the moon and sun gravitational attraction, triggering numerous small (some large) earthquakes daily that shatter rocks by brittle failure. The shattered earth is wrinkled by tectonic deformations and by a myriad of fractures, joints and faults which find morphologic expression at the surface and at depth. Good aerial photographs reveal that surface rock fracturing is a normal phenomenon, as numerous lineations in dominant directions express.

One of the main problems of photogeology is recognizing significant major faults. Well logs, where available, determine the proximity to major faults and cause fracture porosity in brittle rocks susceptible to oil and gas entrapment.

Under rhythmic tidal pulsations and flexing of the earth's crust, deep-seated fractures may develop upward to the surface and even through relatively unconsolidated sediments. Flexing action of terrestrial tides is believed to cause such deformations as the Gulf Coast growth faults (or gravity or contemporaneous faults) with their associated rollover and/or drag along the fault surfaces.

A similar phenomenon has occurred along the Luling-Mexia fault zone within more competent and brittle rocks. In either case, tension stress causes rock failure along major faults, giving rise to porosity development as openings between separated fracture faces. Along this fault zone, competent beds (Austin chalk, Buda lime and Edwards lime) are upthrown coastward and act as canti-lever beams, the flexing of which caused fracture sets of decreasing opening separation, length and frequency of occurrence away from the major faults. Figure 7-3 schematically represents this condition. Fracture intensity measuring computations based on well logs should permit remote sensing of such wells' relative proximity to major faults.

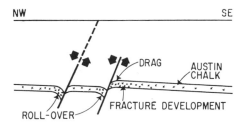

Figure 7-3. Cantilever or "up-to-the-coast" fault system of the Luling-Mex-ia-Talco trend. (Courtesy of *World Oil.*)

Fracture Intensity Index (FII)

An important parameter of fractured and vuggy rocks is the apportioning of total porosity between intergranular porosity (fine pores) and large pores (vugs, fractures, fissures, joints, etc.). For the FII, it is better to determine the "actual fracture porosity" rather than the porosity partitioning coefficient. The total fracture porosity, regardless of the preexisting intergranular porosity, pre-sumably is the factor which quantitatively measures the intensity of brittle rocks' deformational shattering.

The problem of finding and evaluating the natural reservoir fracturing has been publicized in recent years, following the intro-

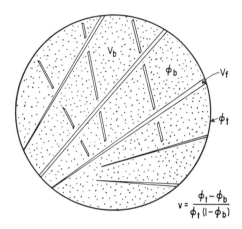

Figure 7-4. Schematic representation of the porosity partitioning between block (or matrix) porosity and fracture-vug porosity. (Courtesy of *World Oil.*)

duction of fracture-finding well logs. Essentially, these techniques utilize the reduction caused by fractures of the amplitude of the acoustic impulse's compressional and shear waves.

Laboratory and field investigations have shown that the relationship between wave amplitude attenuation and fracturing is not simple. Horizontal fractures cause little or no attenuation of the compressional waves. On the other hand, horizontal fractures considerably attenuate shear wave amplitudes. Attenuation for plane compressional and shear waves varies considerably with the fracture plane's dip angle relative to direction of wave propagation. Well bore rugosity strongly affects amplitude reduction. Tool eccentricity and hole enlargement or ellipticity are major factors in reducing sonic amplitude. A similar effect results from gas bubbles in the mud.

These and other complications have made it impossible to quantitatively determine acoustic amplitude reduction in terms of the degree of porosity development that may result from natural fracturing. Also, it is desirable to determine rock fracturing from all types of old open hole well logs—particularly, conventional electrical logs—since running new fracture finding logs in cased wells is impossible.

The following examples briefly reviews techniques for determining fracture porosity from conventional electric logs, Pirson[17] previously discussed.

Figure 7-4[17] shows a hypothetical cross section of a core from a fractured rock.

Let: V_b = total volume of block or matrix

V_f = total volume of fracture-fissure system

ϕ_t = total porosity, including block and fracture porosity

ϕ_b = block or matrix porosity.

Then: Total pore volume=

$$(V_b + V_f)\,\phi_t = V_b\phi_b + V_f \qquad (7\text{-}1)$$

since the fracture-fissure pore space (V_f) is 100 percent pore volume. Let: v = fraction of total pore volume in the fracture-fissure system.

$$v = \frac{V_f}{V_f + V_b\phi_b} \qquad (7\text{-}2)$$

Then: matrix and fracture volumes are eliminated from (7-1) and (7-2), where

$$v = \frac{\phi_t - \phi_b}{(1 - \phi_b)\,\phi_t} \qquad (7\text{-}3)$$

By definition, the FII is the fraction of total porosity developed by fracturing, or $FFI = v\phi_t$.

Therefore:

$$FFI = \frac{\phi_t - \phi_b}{1 - \phi_b} \qquad (7\text{-}4)$$

This index measures porosity development induced by fracturing. Electric logs can determine if these conditions are reasonably satisfied:

1. The mud filtrate displaces all reservoir fluids from the fracture-fissure system within the invaded and flushed zones. Such

zones may not be distinguished from each other in any case when dealing with fractured rocks.

2. The blocks are not flushed by mud filtrate or deeply invaded.

3. Resistivity log curves, one of which probes the invaded zone, exist. This must be done without much geometric distortion of the readings (i.e., from adverse borehole and bed thickness effects). In some cases, such distortions must be corrected.

Those conditions are reasonably met with short and long normal curves at least in chalk reservoir rocks. A short normal and an induction curve may be adequate for other rocks. These equations may be written when logging is at a 100 percent water saturation level in blocks and fractures:

(1) *Short normal equation*

$$\frac{1}{R_{fi}} = \frac{v\phi_t}{R_{mf}} + \frac{(1-v)\,\phi_t}{R_{bo}} \qquad (7\text{-}5)$$

(2) *Long normal or induction log equation*

$$\frac{1}{R_{fo}} = \frac{v\phi_t}{R_w} + \frac{(1-v)\,\phi_t}{R_{bo}} \qquad (7\text{-}6)$$

Equations 7-5 and 7-6, in effect, place two electric circuits in parallel, i.e., the fracture-fissure system filled with mud filtrate (R_{mf}) and the block matrix system (resistivity R_{bo}) filled with formation water (resistivity R_w). R_{fi} and R_{fo} are, respectively, the short normal resistivity (after correction for borehole effects or restoration) and the long normal resistivity (similarly corrected or restored, if necessary), or the induction log resistivity. The "i" means that the reading is from the invaded zone; "o" signifies that the reading is from a zone of 100 percent formation water saturation.

These measurements preferably should be made in a water-bearing fractured rock. In an oil-bearing fractured rock, uncertainties are introduced because of the partial water saturation in the undisturbed zone's fracture-fissure system. The calculated fracture intensity index by 7-5 and 7-6 is, then, too low if oil is not allowed for. However, this is seldom encountered when the FII maps fracture development or determines the approximate distance to a fault.

Figure 7-5 illustrates the calculation procedure with the log section from a producing well in the Salt Flat-Tenney Creek Field, Caldwell County, Texas. The section is from the producing Austin chalk. It appears fractured as observed from the hachured SP curve, each pip of which likely corresponds to a streaming potential effect resulting from mud filtering into the largest fractures.

Basic data pertaining to that level are

d = 7 7/8-inch hole diameter

tf = 111°F formation temperature

R_m = 4.4 @ tf

R_{mf} = 3.5 @ tf

R_{mc} = 5.3 @ tf

R_w = 0.125 @ tf

From log readings at 2,390 feet, R_{fi} = 16.0 and R_{fo} = 8.5. Solving equations 7-5 and 7-6 simultaneously, one finds $v\phi_t$ = FII = 0.0125.

The fault proximity index (FPI) calibration (Figure 7-6)[17] suggests that this well should be approximately 2,000 feet from the fault responsible for the shattering. However, this well is a producer; and the formation water saturation in the fracture-fissure system is approximately 40 percent; the well makes a water cut of approximately 25 percent. Allowing for oil in the fracture system, the FII is computed at 0.028. The indicated distance to the fault on this basis is 200 feet. This well actually cuts the major fault 1,850 feet deep. An up-to-the-coast displacement of 110 feet is in the Taylor section.

The data about double faulting (Trunz[20]) at the Branyon-Buchanan oil field, Caldwell County, Texas, have been plotted on Figure 7-6, providing the fracture intensity index (FII) in the Austin chalk as a function of the horizontal distance in feet from the fault trace. The FII is the total porosity induced by flexing the

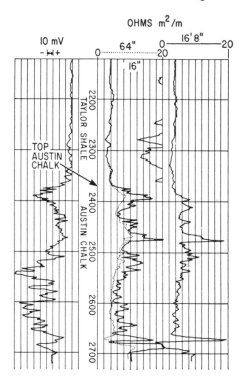

Figure 7-5. Conventional electric log (SP-2 normals-1 lateral) in the fractured Austin chalk, Salt Flat-Tenney Creek fields, Caldwell County, Texas.

Austin chalk, which acted as a cantilever beam anchored and braced to the southeast within the sinking Gulf embayment as it was raised to the northwest in a partially unsupported manner. The double fault indicates partial support.

This uplifting was not without friction, as the drag along the fault surface indicated, or without flexing as the gradually decreasing FII values away from the fault suggested. This formation bending causes lesser regional dips in the upthrown blocks. Conversely, downthrown blocks are subject to more bending because of the rollover effect; hence, the FII does not extend as far laterally for the downthrown blocks (as deformation sensors) as the upthrown blocks do. Using Figure 7-6 to predict distances to faults (applying this relationship only to the general Luling-Mexia fault zone area or to similar rock with similar loading), one must know if a well is on the suspected fault's downthrown or upthrown side. This generally will be known from subsurface geological studies.

The FII may be measured on the object bed if it is susceptible to becoming a fractured reservoir, such as the Austin chalk, the

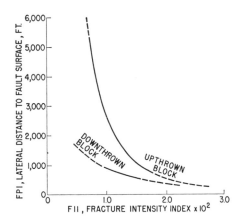

Figure 7-6. Fault Proximity Index (FPI) as a function of Fracture Intensity Index (FII) in the Austin chalk. (Courtesy of *World Oil.*)

Annona chalk (North Louisiana) or the Selma chalk (Mississippi and Alabama along the Gilbertown-Pollard fault trend). Similar studies can be made in fractured reservoirs, not necessarily associated with major faults (i.e., Spraberry sand in West Texas and the Morrow sand in Oklahoma), where deep-seated basement wrenching may have induced fracturing. Fracturing may vary laterally in intensity, according to position with respect to structural deformation. The FII may guide exploration for fractured reservoirs as a mapping index or plan their development after discovery.

The FII, also, may be measured in a brittle formation close to a target oil reservoir which is not necessarily brittle. This exists in

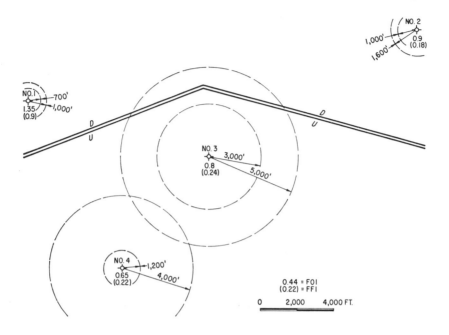

Figure 7-7. Method for determining the probable position of fault traces in the Austin chalk based on log-derived Fracture Intensity Index (FII) and Fault Proximity Index (FPI). (Courtesy of *World Oil.*)

the search for growth faults along the Louisiana and Texas coasts, where the Heterostegina lime is. This formation behaves much like the Austin chalk and occasionally is fractured reservoir. Its fracture intensity index may measure the proximity to growth faults and the shattering they and salt and shale uplifts cause. Mapping the Heterostegina lime's FII may help distinguish truncations from growth faults and diapiric intrusions.

Fault Proximity Index (FPI)

Figure 7-6 suggests using the FII as an exploration parameter. This appears in Figure 7-7,[17] where conventional electric logs were on four wells. General subsurface geology implies that a Luling-Mexia type fault may be expected and that wells 1 and 2 are on the downthrown side. Techniques described herein must not be expected to predict distances to a fault with great accuracy; instead, the calculations should bracket the probable distance to a fault between minimum and maximum distances. These depend on averaging techniques used for reading the minimum and maximum excursions of resistivity curves at the fracture development levels.

Circles with wells as centers represent calculations. The probable fault position, as computations of the FPI minimum and maximum suggest, is on Figure 7-7. This fault trace exhibits the appropriate dogleg presumably required for lateral oil entrapment in these fault line fields: Luling, Branyon, Buchanan, Salt Flat and Darst Creek. All exhibit this lateral entrapment feature.

Aerial photographs of this area show numerous fault lineations with predominant direction parallel to the derived faults. From them it may not be possible to distinguish the major break and minor fractures and joints. Mapping the FPI minimum and maximum may remove an ambiguous aerial photo interpretation and verify the major fault's surface trace.

Fracture Density Index (FDI)

This parameter is not only from logs; it may be derived from any structural map of a brittle bed. Structure mapping only requires determining elevation at the top of a formation.

Gomez[7] has shown with structural deformation models that the greatest probability of fracture occurrence in a brittle bed is at or near the structural surface's maximum curvature. Accordingly, obtaining a structural surface's second space derivative should point out the areas of maximum probability for fracture occurrences, i.e., the FDI. Such an index could not be derived for the four wells in Figure 7-7, as the well control is insufficient; the structure is not appropriate, either. Trend surface analysis best studies the FDI, i.e., by fitting second or third degree structural surfaces to the available structural data and then taking the second derivative.

The theoretical justification of this approach follows:

Consider a low relief structure containing a brittle layer of "T" thickness and of "R" curvature's radius in a plane intersecting this surface and passing through that plane's center of curvature. Let fracture openings develop in that plane at "ds" distance, which defines an angle $(d\alpha = \tan^{-1} \frac{ds}{R})$ between the fracture faces.

Fracture porosity associated with the block of volume = 1 x ds x T is given by

$$\phi_f = \frac{\frac{1}{2} (Td\alpha) \; T \; x \; 1}{1 \; x \; ds \; x \; T} = \frac{1}{2} \; T \frac{d\alpha}{ds} \qquad (7\text{-}7)$$

The radius of the structure's curvature in the object plane is given by

$$R = \frac{ds}{d\alpha} = \frac{\left(1 + \dfrac{dH}{dx}\right)^{3/2}}{\dfrac{d^2 H}{dx^2}} \qquad (7\text{-}8)$$

Where $H = R \tan \alpha$ is the structural height above a reference datum plane passing through the center of curvature. However, for a low relief structure, $\frac{dH}{dx} = 0$ and

$$R = \frac{ds}{d\alpha} \cong \frac{1}{\dfrac{d^2 H}{dx^2}}$$

Therefore, the fracture porosity (ϕ_x) developed in the x direction is given by

$$\phi_x \quad = \frac{T}{2} \; \frac{d^2 H}{dx^2}$$

Similarly, the fracture porosity (ϕ_y) developed in the y direction is given by

$$\phi_y \quad = \frac{T}{2} \; \frac{d^2 H}{dy^2}$$

The total porosity (ϕ_t) the fracturing developed totals the two, yielding

$$\phi_t = \phi_x + \phi_y = \frac{T}{2}\left(\frac{d^2H}{dx^2} + \frac{d^2H}{dy^2}\right) \qquad (7\text{-}9)$$

since ϕ_x and ϕ_y are additive in a warped surface.

With Laplace's equation, without source or sink,

$$\frac{\delta^2H}{\delta x^2} + \frac{\delta^2H}{\delta y^2} + \frac{\delta^2H}{\delta z^2} = 0 \qquad (7\text{-}10)$$

$$\phi_t = -\frac{T}{2}\frac{\delta^2H}{\delta z^2} \qquad (7\text{-}11)$$

Accordingly, a measure or an index of the total porosity that tectonic deformation and fracturing developed may be taken with the vertical second derivative of a structure map drawn on top of the brittle formation of substantially constant "T" thickness.

Techniques for determining and mapping the vertical derivative of a space function are well known. Pirson[15] has applied them; and Peters,[14] Elkins[5] and Henderson[8,9] have presented their principles. Muñoz[13] has specifically applied these techniques to fracture trend analysis in a particular area with respect to oil production in fractured Austin chalk, Frio County, Texas, (Figure 7-8[13]).

The structure map of the area shows the prominent Pearsall anticline, along which an Edwards lime (southwest) accumulation and a separate Olmos sand (northeast) accumulation have been discovered. The contour map shows, in addition, a prominent and less prominent nose axis and a trough axis. The dipping beds north of the Pearshall anticline are, therefore, slightly deformed and warped; it is especially important to delineate large FDI areas in locating Austin chalk wells that would have the greatest chance of containing commercial oil reserves.

Using the structure map in Figure 7-8, an interpolation procedure to grid the readings every 1,000 feet, the Henderson grid in

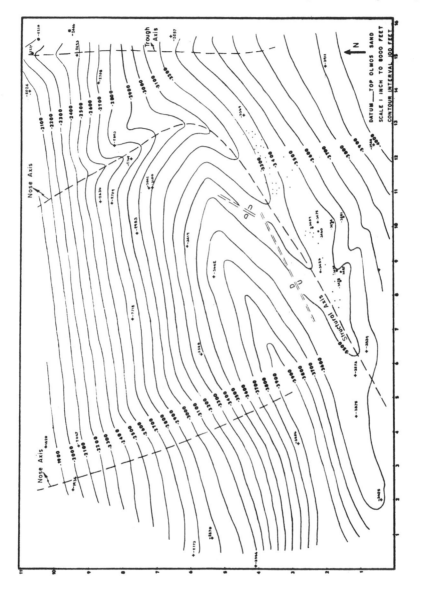

Figure 7-8. Structure map on top of the Olmos sand in the Pearsall Field region, Frio County, Texas.

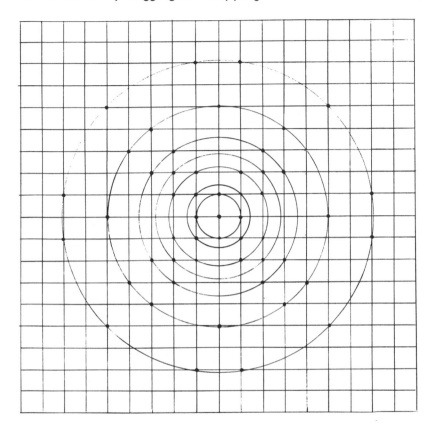

Figure 7-9. Grid mesh point for calculating the vertical second derivative of a function distributed in a horizontal plane. (Courtesy of R. G. Henderson, I. Zietz and *Geophysics.*)

Figure 7-9[9] to compute average readings around its circle and a computer program in Appendix III to eventually summarize the readings, Muñoz obtained the FDI map in Figure 7-10.[13] This map shows some striking trend of high fracturing, and many Austin chalk prospects are so delineated. They await the acid test of the drilling.

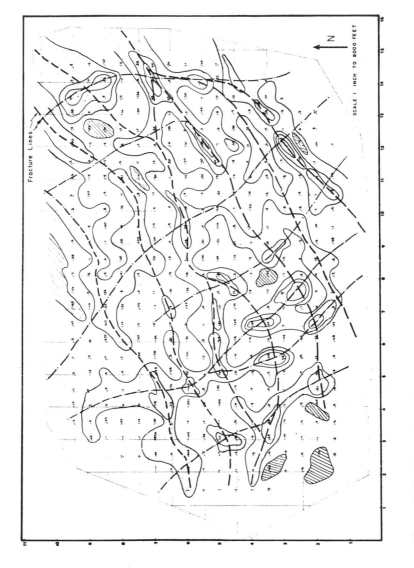

Figure 7-10. Second vertical derivative map of the structure shown in Figure 7-8 and which indicates the expected major fracture trends in the area. (Courtesy of R. E. Muñoz-E.)

Example of Application

The log segment in Figure 7-11 is from Caldwell County, Texas. It shows the Austin chalk section with top at 1,990 feet. It is in the upthrown block of an expected fault similar to that which controls oil and gas accumulation in the Salt Flat and Luling-Banyon fields nearby. The following information about the log calculates the FII and FPI:

Rm = 2.2 Ωm @ 110° F (t_f)

Rmf= 1.8 Ωm @ t_f

Rw = 0.2 Ωm @ t_f (Austin chalk water)

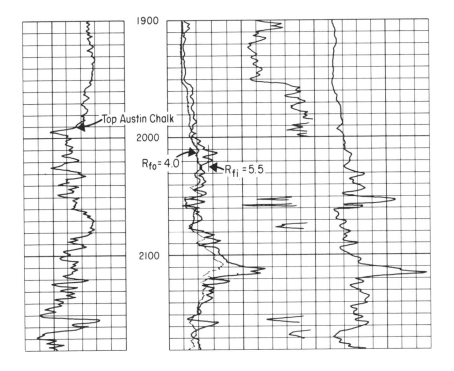

Figure 7-11. Example log for fracture intensity and fault proximity index calculations.

The resistivity curves are from a short normal-induction log combination. In the 2,010-2,020 foot interval, these readings result:

Rfi = 5.5 from the SN curve
Rfo= 4.0 from the IL curve
Combining equations 7-5 and 7-6,

$$FII = \frac{\dfrac{1}{Rfo} - \dfrac{1}{Rfi}}{\dfrac{1}{Rw} - \dfrac{1}{Rmf}} = 0.015$$

At this level is 1.5 percent porosity of bulk rock volume, which fracturing generates. Figure 7-6 shows it is approximately 1,000 feet to the fault.

References

1. Bouma, A. H., "Sedimentology of Some Flysch Deposits," Amsterdam: Elsevier Publishing Company, 1962.

2._____ and A. Brouwer, "Turbidites," *Developments in Sedimentology #3,* Amsterdam: Elsevier Publishing Company, 1964.

3. Bugbee, J. M., "Reservoir Analysis and Geologic Structure," AIME Technical Publication No. 1527, October, 1942.

4. Doyle, W. M., Jr., "Tapping the Austin Chalk's Rich Reserves," *The Oil and Gas Journal,* 54, No. 3, (December 5, 1955), pp. 180-182.

5. Elkins, Thomas A., "The Second Derivative Method of Gravity Interpretation," *Geophysics,* 16, No. 1, (1949), pp. 29-50.

6. Flores, G., "Evidence of Slump Phenomena (Olistostromes) in Areas of Hydrocarbon Exploration in Sicily," Paper No. 13, Section I, Proceedings Fifth World Petroleum Congress, New York, 1959.

7. Gomez, P. L., "An Experimental Study of Fracture Formation in an Anticlinal Structure," Master's thesis, The University of Texas, January, 1967.

8. Henderson, R. G., "A Comprehensive System of Automatic Computation in Magnetic and Gravity Interpretation," *Geophysics*, 25, No. 3, (1960), pp. 569-585.

9. _____ and I. Zietz, "The Upward Continuation of Anomalies of Total Magnetic Intensity Fields," *Geophysics*, 14, No. 4, (1949), pp. 508-516.

10. Koester, E. A. and H. L. Driver, "Symposium on Fractured Reservoir," *AAPG Bulletin*, 37, No. 2, (February, 1953), pp. 201-330.

11. Martin, G. H., "Petrofabric Studies May Find Fracture-Porosity Reservoirs," *World Oil*, 156, No. 2, (February, 1963), pp. 52-54.

12. Moore, C. A., *Handbook of Subsurface Geology*, New York: Harper and Row, 1963.

13. Muñoz-E., R. E., "Fracture Finding by Structural Curvature Mapping," unpublished Master's thesis, The University of Texas, January, 1968.

14. Peters, Leo, "The Direct Approach to Magnetic Interpretation and Its Practical Application," *Geophysics*, 14, No. 5, (1949), pp. 290-320.

15. Pirson, S. J., "Quantitative Interpretation of Gravity-Meter Data," *The Oil Weekly*, 117, No. 7, (April 16, 1945), pp. 34-42.

16. _____ "Comprehensive Quantitative Well Log Interpretation in Multiple-porosity Type Reservoir Rocks," *Handbook of Well Log Analysis*, Englewood Cliffs, New Jersey: Prentice-Hall, Inc., 1963, pp. 303-314.

17. _____, "How to Map Fracture Development from Well Logs," *World Oil*, 164, No. 4, (March, 1967), pp. 106-114.

18. _____ et al., "Fracture Intensity Mapping from Well Logs and from Structure Maps," Paper B, SPWLA Symposium, Denver, June, 1967.

19. Sangree, J. B., "What You Should Know to Analyze Core Fractures," *World Oil,* 168, No. 5, (April, 1969), pp. 69-72.

20. Trunz, J. P., Jr., "Some New Interpretaiton of Reservoir Properties by Well Log Analysis," Master's thesis, The University of Texas, August, 1966.

21. Zemanek, J. V. et al., "The Bore-hole Televiewer," *Journal of Petroleum Technology,* 21, (June 1969), pp. 762-774.

Hydrogeology I:
Hydrodynamics of Compaction

A waterborne process of hydrocarbon migration and accumulation is almost universally accepted as the primary natural way to pool oil and gas in structural and stratigraphic traps. It calls for large volumes of fluid transfer from sedimentation to the present. Field examples postulate and verify that radioactive elements can trace fluid motion throughout geologic time and that significant radioactive mapping patterns are derivable from gamma ray logs, which characterize hydrocarbon accumulations at depth. Hydrocarbons also leak from their loci of entrapment, thereby modifying the physico-chemical properties of overlying sediments—more specifically, their redox potential.

In a sedimentary basin, where rocks were originally laid down in sea water as detrital materials that surface run off brought in or as accretionary carbonates, one of the obvious consequences of induration and lithification is the expulsion during geologic time of most of the water originally in the rocks. Much evidence indicates that the water expulsion occurred laterally and vertically, depending on the relative ease of water movement at various stages of rock consolidation. In certain sediments, large volumes of formation water must have filtered through each unit of rock volume during geologic time. Water movement greatly changed water and rock compositions on a macro and micro scale. This is a branch of the vast field of geochemistry, some aspects of which may be investigated by well logs.

Chapters 8–10 mainly examine paleohydrodynamics, i.e. movement of formation fluid which may have occurred in the *distant geologic past,* causing oil and gas accumulations; or which, in the more *recent geologic past,* have caused flushing by surface waters of infiltration.

Three hydrogeologic periods are considered.

Chapter 8 examines the period in the distant past in which the fluid motion began soon after sediments were laid down and which may have continued for a million to a billion years. The water motion occurred mostly toward land and vertically upward from compaction.

Chapter 9 studies the period when water expulsion under compaction was prevented from occurring at a sufficiently rapid rate to maintain hydrostatic equilibrium and when the overburden pressured up pore waters and when regions of abnormally high pressures were created.

Chapter 10 covers the period in the recent past during which fresh water infiltrated at the outcrops. Water moved toward the basin's center, causing formation water dilution and hydrodynamic entrapment.

Genesis of Geochemical Anomalies

New concepts view geochemical anomalies associated with oil and gas fields as fossil remnants of hydraulic leakage upward from the bottom of sedimentary basins during their formation, compaction, uplifting and erosion. This hydraulic leakage occurred mostly throughout active sedimentation and compaction. Hydrocarbons of all kinds, salts from rocks (in particular, radioactive salts) and trace chemical elements usually associated with oil and gas in generation, migrated upward with the waters in motion. Environmental conditions changed from oxidation to reduction.

Figure 8-1[26] best illustrates genesis of geochemical anomalies over and around oil and gas fields while they were being formed. Here a marine basin is to the left of a shoreline. Detrital sediments

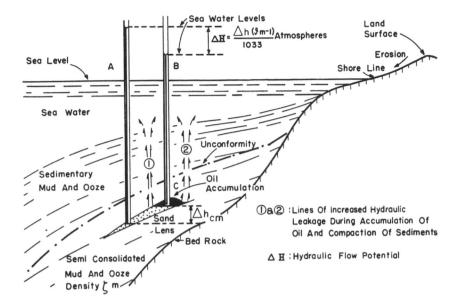

Figure 8-1. Illustration of the development of a stratigraphic trap and attendant hydraulic relationships, including leakage around the edges of the oil deposit as it reaches a mature development stage. (Courtesy of *World Oil.*)

are transported for deposition in the basin, subsiding under sediment pressure. Recent sediment examination (Smith[29]) shows that petroleum-like material is deposited with sediments and ooze in sufficient disseminated volume, to explain all the known oil fields in the world. All that is required to form an oil field is migration through concentration and accumulation in suitable geologic traps.

Consider such a trap to be a buried sand bar or "C" lens of relatively clean sand relative to the mud and ooze that enclose it. Now visualize manometers A and B (i.e., well casing) embedded in the sediments but terminated at their bottom in the clean sand lens. In each manometer clear salt water rises to equilibrium with hydrostatic and geostatic (overburden) pressures at the respective bottom of each well. The water level will be higher at manometer A than in manometer B because of the greater ooze and mud thickness around well A and, particularly, because of the uncon-solidated nature of the sediments and their incompetency.

According to Darcy's law of fluid motion in porous media, a hydraulic flow of formation water will occur from A to B within the sand lens. In so doing, this fluid contains countless tiny oil globules. At the top of the lens, these globules cannot escape into the overlying ooze and mud because of their fine capillary size. The oil droplets are filtered out at the upper end of the lens because oil is the nonwetting phase and would have to overcome a large capillary pressure gradient to re-enter the mud and ooze, which eventually will become shale.

Hence, oil and gas accumulate at the top of the sand lens and continue as long as the hydraulic flow is sustained. However, permeability to water no longer exists in the accumulated oil zone, and hydraulic leakage is forced around the accumulation (Figure 8-1). This oil accumulation may be a revival of the Munn theory.[22] It appears as a very realistic explanation not only of oil and gas accumulation in stratigraphic and structural traps but also of mineralization effects in unconsolidated sediments overlying developing oil and gas accumulations.

As formation water migrates around the petroleum deposit from greater to shallower depths, it cools and precipitates salts within the geologic section, causing a zonal distribution around the oil field. Absorption of salts, ions and hydrocarbons may also play a part in this mineralization.

One of the most striking evidences of the vertical formation water leakage during sediment consolidation and compaction is the earth's low resistivity over salt domes even when buried more than 1,000 feet below the investigation depth of surface resistivity measurements. Presumably, upward salt water migration around salt domes rendered the shallow formation waters more saline, thus resulting in lower resistivities. Further evidence of upward water migration, partly in vapor form, lies in the higher salinities in a geologic horizon at great depths, compared to lower salinities at shallow depths in the same horizon. Arbuckle waters attain 300,000 ppm salt content in the Anadarko Basin in Oklahoma 10,000 to 15,000 feet deep; in northeastern Kansas, Arbuckle water salinities of 2,000 to 3,000 ppm are approximately 2,000 feet deep.[13]

Flerov[15] has shown that radioactive anomalies may carry to substantial depth by investigating the composite radioactivity of wells in and around the Guselky oil field. He observed that the radioactivity of cores from producing wells was smaller than from wells immediately close to the edge of the field.

Radioactivity Anomalies

Currently accepted views on the origin of hydrocarbons in the earth leave little doubt that the oil-gas migration and pooling in geologic traps was a waterborne process. The detail mechanism of oil origin, migration and entrapment is not clearly established. Americans like Smith,[29] Baker[7] and Meinschein[21] visualize the process as one of selective solubilization in colloidal soapy water suspensions and sieving out of the hydrocarbons at the entrapment locus. Canadians like Hodgson[16] visualize oil migration as accomodating hydrocarbons in water and salting out at accumulation. Both systems find it difficult to explain the origin of the light and asphaltic fractions of crude oils. Frenchmen (i.e., Bordenave[11]) remedy the situation with low temperature pyrolization of organic substances, which generates the full spectrum of hydrocarbons and the relative concentration of each, which depends on many geologic, catalytic, composition, duration and temperature factors.

The above theories claim that the end mechanism of accumulation is a waterborne process of transportation and of collection over geologic time into a *primary* entrapment locus. (Here we are not concerned with possible *secondary* migration of oil already pooled into a new entrapment position.)

Whatever the release mechanism may be, another important point is what happens to the soapy waters after they have released their hydrocarbon load. Since they are colloidal suspensions, they migrate through compacting sediments of the marine basin of accumulation. Buoyancy causes migration through the shales and ooze of the as yet nonlithfied sediments, which can be only in one main direction—namely, vertically upward and away from the gravity forces (Figure 8-2[26]). Local but important lateral motion of the compaction expelled waters may occur when the beds are already stratified and tilted. It is generally safe to assume that the expelled water and its solubilizers move vertically upward.

Obstacles of varying magnitudes will naturally block the vertical paths of these waters, especially in porous formations where other oil accumulations may already have occurred or in adsorption barriers to the water flow that fine grain sediments, may have provided, or permeable beds may carry them laterally. Because of restrictions to the vertical expulsion of formation waters, soap micelles tend to form adsorption anomalies at or within such restrictions. Anomalies exist at all such partial adsorption barriers as long as the oil continues to accumulate and will be found throughout the geologic column as long as sedimentation continues during oil and gas migration and as long as they are not transported laterally. The effect of oil at depth may thus be sensed from thousands of feet above the actual accumulation by detecting the anomalous distribution of mineralization by salts of organic acids in the geologic column.

Depending on the pattern of *barrier* and *carrier* beds interposed between the developing oil field and the earth's surface, various organic mineralizations of overlying sediments are possible. Mineralization anomalies are in shales, but may also be in shaly sands, carbonates and massive anhydrite beds. Shaly water sands may also cause mineralization anomalies.

Active migration and accumulation of solubilized hydrocarbons must cease eventually when removable oil droplets of the carrier beds and associated source beds have been selectively extracted. This, however, does not terminate water expulsion from compacting sediments still undergoing lithification in a sedimentation basin. Oil accumulation may also terminate from exhaustion of the solubilizers; this is possibly the main reason for an oil and gas accumulation's stopping.

Over geologic time, the sedimentary basin continues to lose its formation waters (now devoid of solubilizers) during subsequent

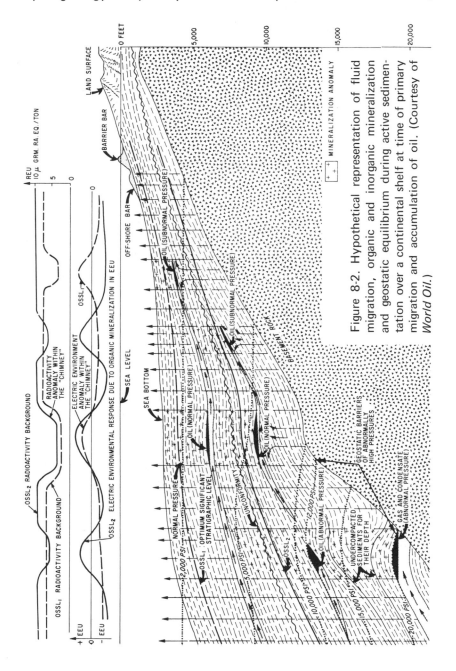

Figure 8-2. Hypothetical representation of fluid migration, organic and inorganic mineralization and geostatic equilibrium during active sedimentation over a continental shelf at time of primary migration and accumulation of oil. (Courtesy of *World Oil*.)

subsidence, while additional deposition of shallower sediments (with or without interruptions, or unconformities) occurs and compaction and lithification proceed. Near oil and gas accumulations, this expelled water must be deflected from its otherwise and normally vertical path, and the water leakage pattern forms a mineralization "chimney" on the boundaries of such accumulations. While such waters now contain almost only inorganic salts and are practically devoid of organic matter, they mineralize the sediment sequences in their upward migration for two main reasons: first, because of earth temperatures' dropping at shallower depths; and second, because of selective adsorption on and desorption of ions from clays and other finely divided minerals.

Other physicochemical phenomena may also exist as a result of the previous organic mineralization (described above). Ions of metallic and metalloid elements are thus selectively retained in a zonal cylindrical manner with mineralization at a minimum level within the vertical cylinder generated on the boundary of the buried oil and gas accumulations. Radioactive elements cause the most readily detected inorganic mineralization. Changes in radioactive mineralization naturally, is most notable in shales. Iodine, bromine, boron and chlorine ions are also adsorbed by clays, and their anomalous distributions within sediments may be used as remote sensing mapping parameters of a petroliferous environment at depth.

Anomalies in mineralization by radioactive elements are readily detected by gamma ray logs properly calibrated in absolute radiation units and by applying proper correction for bed thickness, borehole effects, mud density, casing, tubing, cement, time constant, logging speed and facies variations. Alekseyev[2] et al. have also observed anomalous mineralizations in other heavy metallic ions—vanadium, manganese, nickel, chromium, cobalt, etc.—by chemically analyzing core samples. Specific neutron activation logs could possibly be devised to provide continuous records with depth of significant chemical elements.

The myriad of liberated hydrocarbon droplets, which have coalesced and accumulated at mechanical or electric sieving barriers, then block the path of the compaction expelled formation waters. These barriers do not necessarily coincide with geologic structures or stratigraphic barriers, although such geologic features are likely to coincide with them. The expelled waters are, to a large extent, diverted to a vertical path around the growing oil and gas pool; and a vertical cylinder of hydraulic leakage may be visualized around the growing oil field, the diameter of which is constantly expanding as the oil field expands. In an oil field's mature development stage, the (above) visualized vertical hydraulic leakage may carry little solubilizer, but its inorganic mineralizing effect continues over geologic time. Such mineralization provides remote inorganic oil field sensors, which are mostly detected by well log response to heavy chemical elements that clays readily adsorbed; these well logs measure the natural radioactivity of sediments.

The genesis of oil and gas fields is thus visualized geologically as an early process which occurred soon after sediments were laid out at sea. Sediment compaction and lithification, however, may continue long after a hydrocarbon accumulation is virtually terminated. Then, the continuing and predominantly vertical flux of the expelled formation water acts on the overlying sediments, as within an ion exchange chromatographic column, replacing loosely held inorganic ions of low molecular weight on clays with more tightly held heavy ions—in particular, with those of radioactive elements: potassium, thorium, uranium, radium. This results in a relatively constant and high level radiation background in overlying shales between oil fields, whereas such mineralization occurred in shales directly over the pools only for reduced geologic time, and their radioactivity is below background level.

Usually, radioactivity of shales directly overlying oil and gas fields is reduced to about half of the background radioactivity between fields. While this working hypothesis appears plausible, it is by no means the only one advocated.

The Russian investigator Alekseyev[1] believes that the low radioactivity overlying oil fields results from base exchange, whereby the radioactive elements are leached out by the organic water soluble compounds escaping from the oil accumulation. (The writer does not consider this a likely process.)

Another Russian investigator, Shneyerson[27], believes that the upward diffusion of organic compounds and hydrocarbons from the oil accumulation renders the overlying sediments oil wet, thereby changing surface properties of the overlying rocks and reducing their adsorptive properties for ions of heavy elements, including radioactive elements. This hypothesis received some support by observing that the flocculation rate of overlying fine sediments is smaller than that of similar sediments in nonpetroliferous areas. (It appears that better physicochemical tests could be devised to verify this mechanism.)

However, the Shneyerson's and Pirson's theories substantially agree on the fundamental mechanism responsible for the radioactive mineralization anomaly. No magic is between petroliferous environments and radioactivity; radioactive elements are merely tracers which record areal variations in the cumulative vertical flux of formation fluids as the overburden expels them. One must not seek a causal relationship between radioactive energy and oil genesis; considerable research effort[36] has proved this hypothesis invalid.

Primary waterborne migration originally of sea water (of substantially constant chemical composition over geologic time), contained radioactive elements in solution as follows:

U	15,000 to 20,000	micro gram/Ton	
Ra	5	micro gram/Ton	(When in equilibrium
Th	500	micro gram/Ton	with U)
K^{40}	420,000	micro gram/Ton	

Oil field waters, often much more saline than sea waters, lack uranium—containing much less uranium than do the original sea waters. Yet, oil field waters are highly radioactive, much more so than "radium waters" from "therapeutic" mineral water establishments. This is due to radium salts. Implications are that uranium from the original formation water has been retained as uranoorganic compounds in the reservoir crude oils. Yet, such compounds decay into radium, which readily forms water soluble compounds and accumulates in oil field waters, especially because oil field waters lack sulfates which otherwise would precipitate radium.

Over geologic time, migrating waters carried their hydrocarbon loads to the entrapment loci, where they were stripped of their uranium ions and hydrocarbons; after unloading them, the uranium poor waters escaped through the caprock and overlying formations. After oil accumulation ceased, water migration did not necessarily cease, as continuing compaction showed; and the vertical escape of additional water caused uranium mineralization to continue around the oil accumulation, leaving a "chimney" or "funnel" of reduced radioactivity immediately overlying the oil pool. Often, it extended throughout the geologic column, thus permitting its observation and detection in overlying stratigraphic shale beds.

The required characteristics of such shale beds include *uniformity of facies and thickness and correlatability and identifiability over large areas*. These disturbing factors may possibly exist:

(1) Highly permeable aquifier with strong lateral water motion overlying the oil accumulation, which would cut off the chimney's vertical extension into shallow horizons;

(2) Shale becoming more carbonaceous, calcareous or argilaceous;

(3) Shale becoming thicker or thinner;

(4) Shale becoming coarser or finer in grain size (then, requiring normalization procedures to reduce the radioactivity measured from the gamma ray curve to a common shale lithologic basis);

(5) Disturbance by abnormal concentration in potassium, especially in evaporite regions;

(6) Lateral merging of effects from closely spaced oil accumulations;

(7) Vertical merging of superimposed accumulations.

Unconformities generally do not disturb the observation of a radioactive chimney, as they do not block compaction or water expulsion.

Recognizing a well substantiated radioactivity low within an appropriate shaly marker bed or "optimum significant stratigraphic level" appears to unquestionably indicate primary oil accumulation at depth. Presently, this has appeared over more than 30 known oil fields in France, Canada and the U.S. While this may not be sufficient cases to establish an absolute cause-and-effect relationship, it must be considered as an unquestionable association under favorable conditions. This association is not expected to hold over secondary oil accumulations resulting from mass migration of oil already pooled. Moreover, the radioactive low or "funnel" is not expected to disappear from the region overlying a pool from which the oil has migrated as a result of later diastrophism. This is because the halflife of uranium, the significant tracer element, is more than 4 billion years.

The radioactive funnel extending to the earth's surface and being detected by surface measurements depend on fulfilling certain requirements (described below).

Disturbances by radioactive fallout, cosmic rays and other atmospheric effects should be eliminated. Airborne and near surface measurements are thus eliminated as operative procedures. A satisfactory technique requires measuring in shallow boreholes or from soil samples which must satisfy the following requirements:

1. Sampling must be in a uniform rock or soil formation susceptible of retaining its radioactive elements even under strong leaching activity by surface waters.

2. Near surface rocks or soils used must be formed in place by weathering of bedrock (autochthonous). Samples must not be taken from valley and channel fills, alluvium, glacial drift, beaches, etcetera.

3. Readings must be normalized to a constant soil type, or clay content, etcetera. A single channel spectrometer should do the measuring; its window should be set to B_i^{214}.

When these conditions are met, the radioactivity low may be found on the earth surface and mapped. That surface measurements have observed radioactivity "lows" is unquestioned. Russian and French investigators,[31] too, have confirmed that the "lows" have also been found at depth from well logs; but, there is less unanimity on their explanation. Lundberg's explanation[20] agrees much with the above views, whereas others are without scientific basis.

However, the Russian geophysicist Alekseyev,[4] who discovered such anomalies from logs at depth and above oil fields, denies any genetic association between the two observations. He claims that the association results from granulometric properties (average sediment grain size) varying over anticlines and structures during geologic time. The evidence for this is not convincing, as the grain size variation of soils over structure is not shown. Alekseyev, however, has studied radioactivity's variation with grain size in detail; normalization of soil radioactivity readings (W. T. Willis, "A Validity Test of the Radioactivity Method of Geochemical Prospecting," unpublished Master's thesis, The University of Texas, January, 1961) were based on his study.

In addition, Alekseyev shows four oil bearing anticlinal structures and one non-oil bearing structure with a good radioactivity chimney. He does not prove that the structure has been tested to the basement rock nor into the basement, for a sharp structure of the type shown could overlie a fractured basement rock susceptible of being a reservoir. A by-passed oil accumulation possibly may be in the "barren anticline." In the Russian practice of formation evaluation, lateral curves which could readily overlook thin oil sands (especially, when shaly) are ineffectively combined. A primary oil accumulation also may have existed in past geologic time and had moved away en masse from later diastraphism, leaving its radioactivity imprint as a fossil "radioactivity chimney" overlying its primary residence.

A novel exploration technique like this incorporates two ways to achieve drill exposure and establish a "discovery batting average."

1. By discovering sufficient prospects and promoting their drilling. (Pirson has found the prospects, but their promotion has proved impossible.)

2. By surveying an active exploratory area and observing the success of later drilling although guided by other techniques.

Pirson did this in 1963 over Upton County, Texas, then using available radioactivity logs from dry wildcat wells and a few from known oil fields as control. Only radioactivity maps of the Woodford shale and of the lower Silurian shale were made, from which the discovery predictions proved amazingly accurate with respect to test wells drilled into the Devonian, the Fusselman and the Ellenburger. This high accuracy does not prove the radioactivity low is genetically associated with oil entrapment nor (by Alekseyev's contention) that the association is structural.

Mapping over the Aneth Field, where the accumulation is partly stratigraphic, only partially showed a relation of radioactivity anomalies to stratigraphic oil entrapment. The Aneth accumulations apparently were in several superimposed (and, at times, somewhat vertically offset) reservoirs, resulting in radioactivity anomalies merging in an otherwise not too suitable overlying shale (Paradox shale) to make the measurements. From the Aneth study, it was concluded, however, that stratigraphic oil entrapment should cause the radioactive chimney effect in the overlying sediments.

Pirson also carried out radioactivity mapping over some strictly stratigraphic oil accumulations in the Mission Canyon formation (Williston Basin). By measuring radioactivity intensity in the overlying Spearfish formation, some perfect anamalous lows were mapped over the Glenburn and Pratt fields, North Dakota. The anomalies could not possibly be associated with a structural granulometric anomaly, as the fields are totally devoid of structural relief. These purely stratigraphic anomalies are in Figures 8-12 and 8-13.

Results of some radioactivity mapping surveys are reviewed briefly:

The first survey was over the Coulommes Field, France.[26] The structure maps are in Figures 8-3,[26] and 8-4,[26] showing the radioactivity of the Callovian marl, a formation that immediately overlies the Bathonian reservoir. The correspondence between the two is striking, to say the least. In addition, well BE$_3$ to the northwest, which had been abandoned temporarily during the 1961 survey, was reopened one year later, flowing at 120 BOPD.

Three surveys were then made in Canada over, respectively, the Innisfail Field (Figure 8-5 and 8-6[26]), the Wizard Lake Field (Figures 8-7 and 8-8[26]) and the Westerose Field (Figures 8-9 and 8-10[25]). In the latter, radioactivity was mapped in various shale horizons overlying the oil accumulation; the "chimney" or "funnel" effect moved upward into shallow horizons more than 6,000 feet, traversing two major unconformities: the Exshaw and the Blairmore.

Another similar study was over the Wellman Field, Terry County, Texas, a Wolfcamp reef structure, (Figure 8-11 [26]). In this particular case, the "funnel" effect shows a transverse disturbance from the northeast with decreasing depth magnitude. The cause of this disturbance could not be established; but, it is thought to be particularly important in the Dewey Lake formation because of erratic and anomalous mineralizations in potassium salts in evaporite basins.

Hydrocarbon Leakage

One tenet of geochemical prospecting for hydrocarbons is that wherever they accumulate they leak out at a slow but continuous rate through the overlying sediments. In the early days of these methods, Pirson[23] measured such microleakage. However, the early measurement techniques greatly lacked accuracy and reproducibility.

With chromatographic microgas analysis, positive separation and identification of hydrocarbons in soil air became possible. Such tests were made over the Hilbig Field, Bastrop County, Texas, (Figure 8-14). This field was selected because of its nearness, its small areal extent of the producing field and because it was repressured by gas injection soon after discovery, continuing to be repressured to the present. The Hilbig reservoir is a serpentine laccolith at 2,280 to 2,795 feet deep from the surface; its average pay thickness is 325 feet. The producing area is estimated at not more than 80 acres, and to 1968 it produced more than 5 million barrels. The field survey was in the summer of 1961 when the pressure was 600 psi, whereas the original pressure at discovery in 1933 was 1200 psi. Thus, if hydrocarbon leakage through the overburden ever occurred, it should still have been operating during the survey.

(Text continued on page 233)

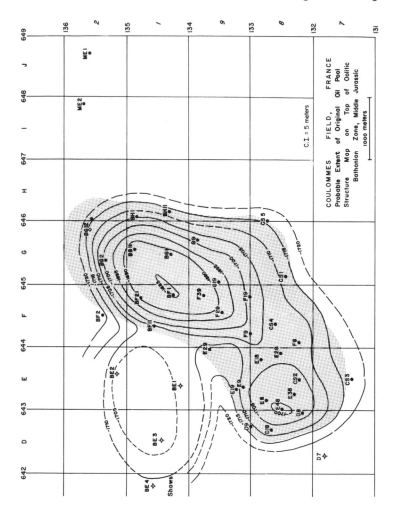

Figure 8-3. Coulommes Field, France. Probable extent or original oil pool structure map on top of oolitic Bathonian zone, Middle Jurassic. (Courtesy of *World Oil.*)

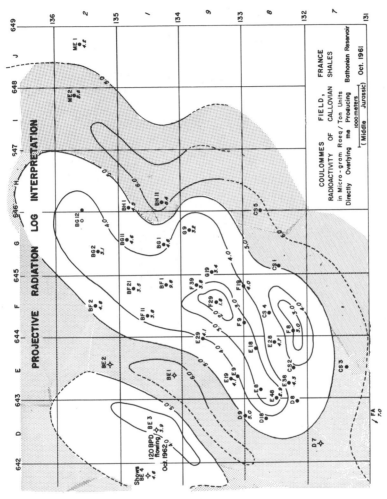

Figure 8-4. Coulommes Field, France. Radioactivity of Callovian shales in microgram Raeq/ton units directly overlying the producing Bathonian reservoir. (Courtesy of *World Oil*.)

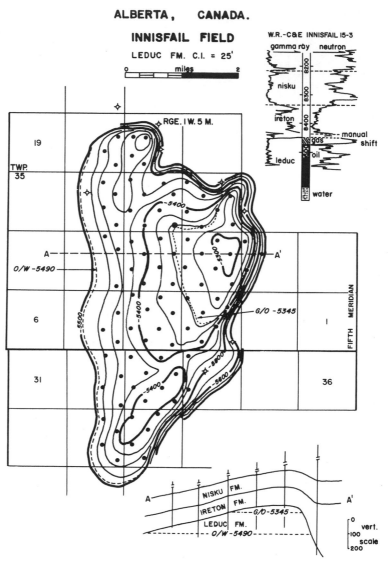

Figure 8-5. Structural map on the Innisfail Field, Alberta, Canada, contoured on top of the Leduc reef. (Courtesy of *The Alberta Society of Petroleum Geologists.*)

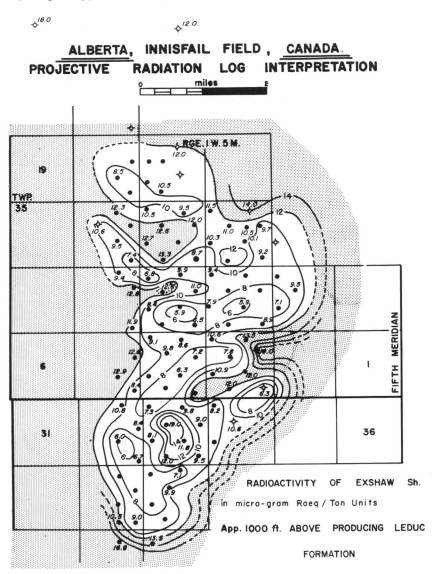

Figure 8-6. Radioactivity of the Exshaw shale in microgram Raeq/ton units app. 1,000 feet above the producing Leduc formation in the Innisfail Field, Alberta, Canada. (Courtesy of *World Oil.*)

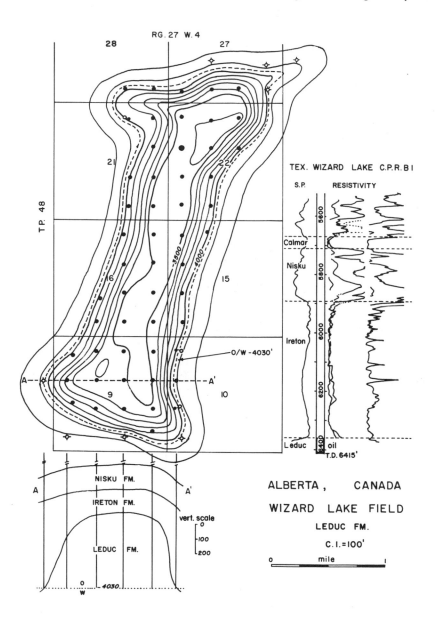

Figure 8-7. Structural map of the Wizard Lake Field, Alberta, Canada, contoured on top of the Leduc reef. (Courtesy of *The Alberta Society of Petroleum Geologists.*)

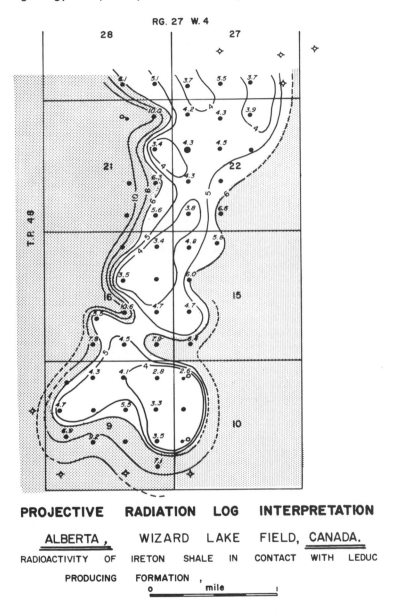

PROJECTIVE RADIATION LOG INTERPRETATION

ALBERTA ,   WIZARD   LAKE   FIELD, CANADA.

RADIOACTIVITY   OF   IRETON   SHALE   IN   CONTACT   WITH   LEDUC

PRODUCING   FORMATION ,

Figure 8-8. Radioactivity of Ireton shale in contact with the Leduc producing formation in microgram Raeq/ton units. Wizard Lake Field, Alberta, Canada. (Courtesy of *World Oil.*)

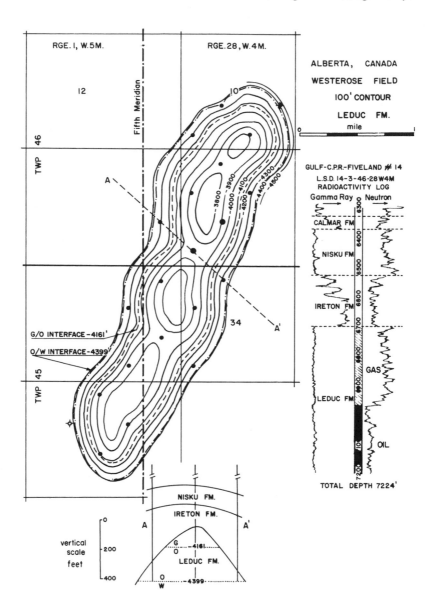

Figure 8-9. Structural map of the Westerose Field, Alberta, Canada, contoured on top of the Leduc reef. (Courtesy of *The Alberta Society of Petroleum Geologists.*)

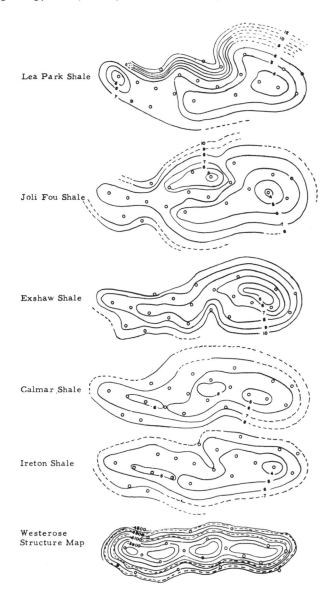

Figure 8-10. Radioactivity anomalies in various shales overlying the Westerose Field, Alberta, Canada. Radioactivity in microgram Raeq/ton units. Note the chimney or funnel effect. (Courtesy of N. Alparone, A. Avadisian and SPWLA.)

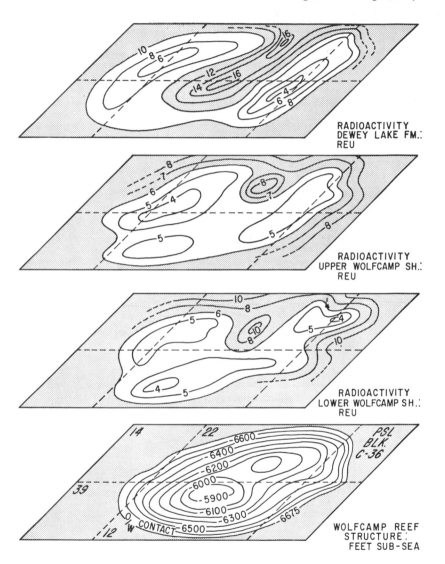

Figure 8-11. Three-dimensional drawing contoured in radioactivity environmental units from gamma ray logs to show patterns at three levels above Wolfcamp reef oil production at Wellman Field, Terry County, Texas. (Courtesy of *World Oil.*)

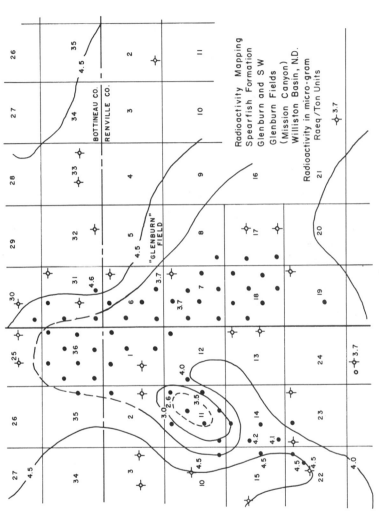

Figure 8-12. Radioactivity mapping of the Spearfish formation. Glenburn and southwest Glenburn fields. (Mission Canyon, Williston Basin, N. D.) Radioactivity in microgram Raeq/ton units.

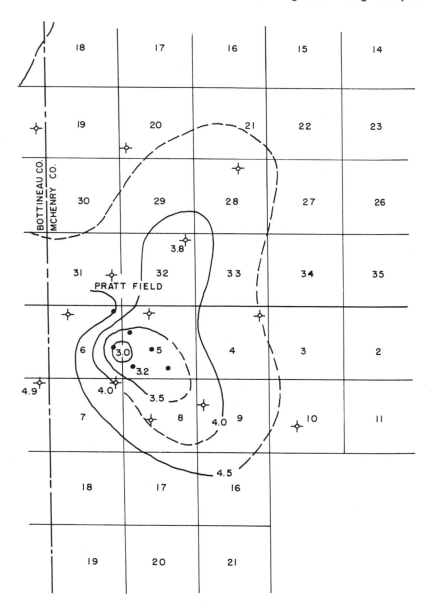

Figure 8-13. Radioactivity mapping of the Spearfish formation. (Pratt Field, Mission Canyon, Williston Basin, N. D.) Radioactivity in microgram Raeq/ton units.

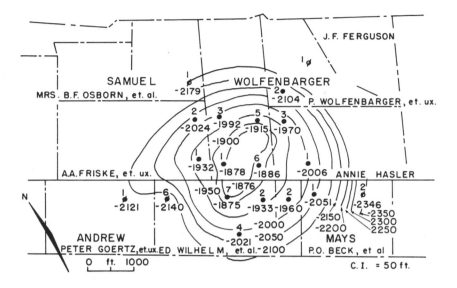

Figure 8-14. Structural map of the Hilbig Field, Bastrop County, Texas, contoured on top of the Serpentine reservoir rock. (Courtesy of Blackburn and AAPG.)

Measurements at 39 stations were distributed more or less uniformly over the field, many of which were beyond the limits of production. Soil and soil air samples were collected from bore-holes 3 to 4 feet deep. A redox probe consisting of a calomel electrode, platinum electrode and glass electrode measured the soil's oxidation-reduction potential and allowed for pH variations in the boreholes. Soil samples were taken to the laboratory for total gamma radioactivity measurements and for normalization by sedimentation technique to a sample of constant specific surface. A chromatograph analyzed the soil air samples for trace hydro-carbons—methane, ethane and propane. Results of the surveys are in Figures 8-15, 8-16, 8-17, 8-18 and 8-19.

Maps of Figures 8-15,[9] 8-16[9] and 8-17[9] give the soil air concentration in the hydrocarbons in parts per million by volume. The high hydrocarbon contents and the limits of production from the serpentine reservoir highly correspond. The redox potential survey of Figure 8-18[9] also indicates satisfactorily the limits of production. Results of the radioactivity of soil samples, (Figure 8-19[9]), leave much to be desired, although one may observe the expected radioactivity reduction immediately over the oil productive acreage. The outline of the accumulation is disconcerting because it is in no way as precise nor as reliable as it is from the hydrocarbon and redox surveys. In fact, the redox survey appears so satisfactory that one might dispense with the elaborate and tedious measurements of soil hydrocarbons altogether in commercial surveys—especially, as hydrocarbon in the soil causes its low redox potential. Alternating the flow of air and natural gas in a 3-foot column of soil in which a redox probe was inserted ascertained this.

Results of some of these experiments are in Figure 8-20. Natural gas passing through this column from June 18 to July 1, 1961, lowered the redox potential of the soil sample by 550 millivolts. When air was again introduced, the Eh potential rapidly returned to its original value and higher. It is expected, as many field measurements have verified, that this phenomenon exists over oil and gas fields from micro hydrocarbon leakage. However, other phenomena—namely, vertical expulsion of oil field waters and telluric currents associated with oil and gas accumulations and other mineral deposits—add to the redox potential changes.

Oil Entrapment by Hydrodynamics of Compaction

Similar phenomena of tilted hydrocarbon water contacts occur when the water flow is in the opporite direction of infiltration, i.e., from the center of the sedimentation basin toward the basin edges. In this case, the salt water flow must be considered.

This phenomenon should be remembered when studying oil and gas accumulations offshore, where many of the low structural relief prospects may be barren because they are flushed by compaction expelled waters (i.e., as many North Sea seismic structures are). In this instance, the objectives are the reservoirs in the Rotliegend sand (Permian) at 6,000 to 7,000 feet deep. Regions of abnormally high pressures reportedly have been in the North

(Text continued on page 240.)

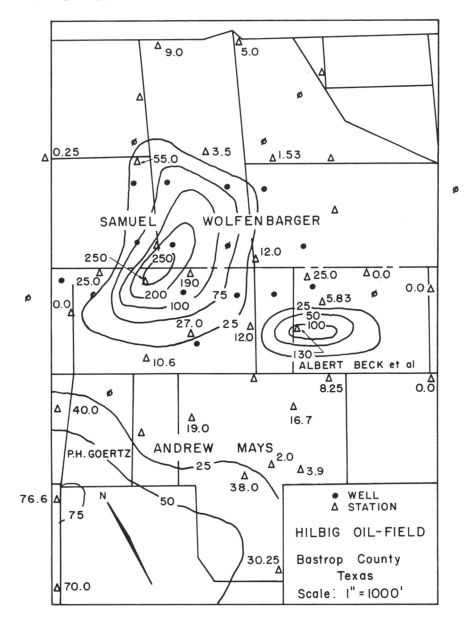

Figure 8-15. Distribution of methane gas in soil air over the Hilbig Field (ppm by volume). (Courtesy of S. Baijal.)

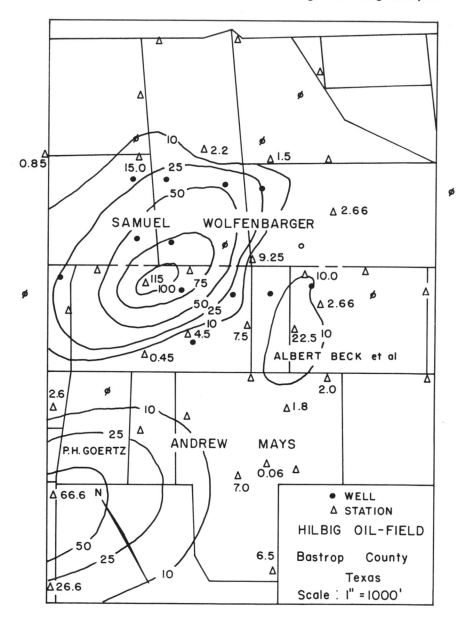

Figure 8-16. Distribution of ethane gas in soil air over the Hilbig Field (ppm by volume). (Courtesy of S. Baijal.)

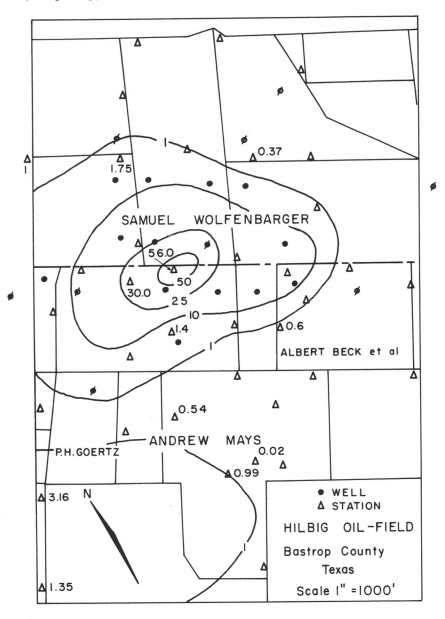

Figure 8-17. Distribution of propane gas in soil air over the Hilbig Field (ppm by volume). Courtesy of S. Baijal.)

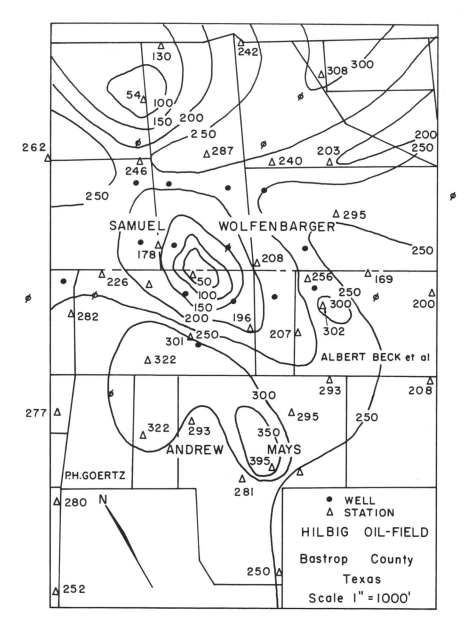

Figure 8-18. Distribution of (Eh) reduction-oxidation potential of the soil over the Hilbig Field (in millivolts). (Courtesy of S. Baijal.)

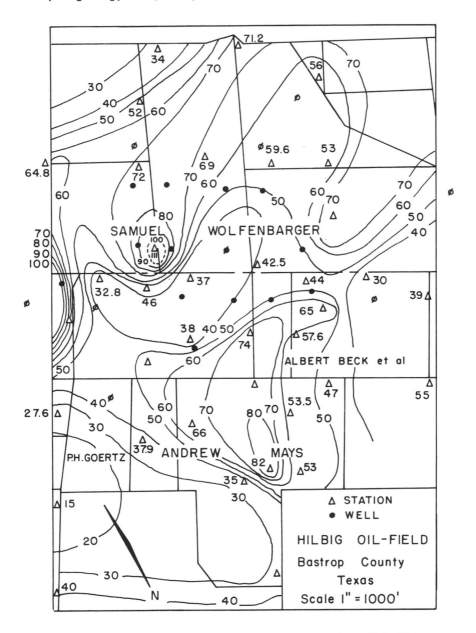

Figure 8-19. Distribution of soil radioactivity over the Hilbig Field (net counts per ten minutes). (Courtesy of S. Baijal.)

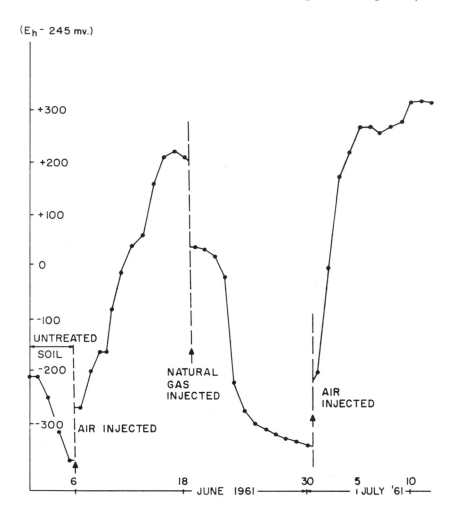

Figure 8-20. Changes caused in the (Eh) redox potential of soil by alternating natural gas and air flow through a soil sample. (Courtesy of S. Baijal.)

Dutch and German parts of the Rotliegend formation, whereas gas fields in the British sector of the North Sea are at normal pressure. A constant southward drift of Rotliegend waters, which has flushed out most of the low relief structures' gas (and oil) when found in the path of the expelled waters, must thus exist.

Radioactivity Mapping Techniques

Given an area to be mapped, various horizons should be selected on logs that show correlatable fine-grained sediments (shales, silts, etc.) that are continuous and that do not vary too much in thickness. In other words, they should be blanket-like, and the survey aims to map the vertical flux of expelled waters through that blanket over geologic time.

On the selected horizons, gamma ray intensity measurements are made by averaging deflections over the selected intervals. Readings are made in the calibration units recorded on the log's heading. (For older logs (pre-1960), various units were used.) Many logs that have no zeros or calibration are also found. Communication with the original service logging company may then be necessary in order to ascertain this information.

Because of the variability of hole conditions, all gamma ray intensity readings must be normalized to a common set of conditions. While corrections may be made in arriving at a true radiation intensity, it is best to avoid making corrections at all levels by selecting as a "normal" well the one most often encountered as to size, casing, mud weight, etc. and to reduce the readings from other wells to the selected normal well's conditions. This is because any correcting procedure is always subject to possible error. Therefore, it is best to avoid making corrections as much as possible.

In the normalization to a standard well, two main cases will be encountered for *thick beds* and for *thin beds*.

Thick Bed Normalization

A thick bed may be defined as one which is 10 to 20 times the length of the radiation detector. For modern logging tools, this is 5 feet. Then, it is not necessary to make a correction for the bed thickness and charts in Figures 8-21a and 8-21b may be used. An (uncased) open hole uses the chart in Figure 8-21a. Its directions are on the chart and are self explanatory. A cased hole uses the chart in Figure 8-21b according to similar explanations written on the chart.

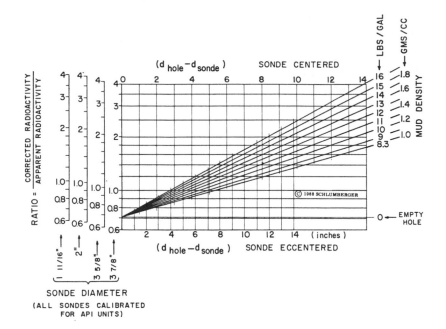

Figure 8-21,a. Gamma ray borehole correction chart for open hole. (Courtesy of Schlumberger Well Services.)

In each, the answer to the correction problem is in terms of the ratio of the true (corrected) radioactivity to apparent radioactivity. If a standard hole has been selected for normalization, the correction factor to that standard hole is from the ratio of the two correction factors, or ratio object well/ratio standard well.

Thin Bed Normalization

When radioactive beds are thin, in addition to the above corrections, the logging speed and the time constant must be allowed for. Both data are on the logs' heading. Logging speed times (x) the time constant gives the "drag" in feet. The drag is actually the travel distance the instrument requires in order to reach nearly full deflection. Figure 8-22 is a chart which may do this by being entered at the bottom with the ratio (bed thickness/drag) and by reading the ratio (true radioactivity/apparent radioactivity) on the chart's ordinate.

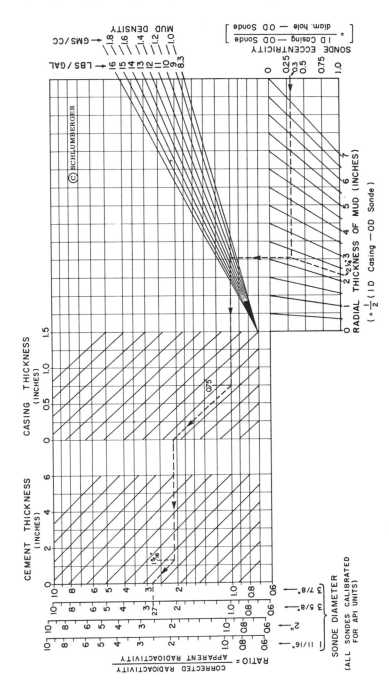

Figure 8-21,b. Gamma ray borehole correction chart for cased hole. (Courtesy of Schlumberger Well Services.)

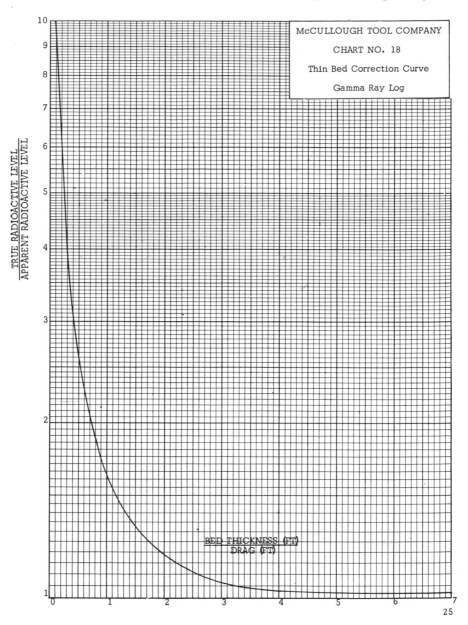

Figure 8-22. Gamma ray thin bed correction chart. (Courtesy of McCullough Services—Baroid Division.)

Again, if a standard hole had been selected for normalization, the correction factor to a standard hole may be obtained by making a ratio of correcting factors.

Reduction to a Constant Radioactivity Unit

For readings to be comparable, they must be expressed in the same unit. The writer actually prefers to use "microgram radium equivalent per metric ton" as the unit because it gives a better idea of the objective—namely, finding and mapping areas of variable mineralization in radioactive elements. However, valid contour maps may be made in API units, mR°/hr, counts per minute, etcetera.

Table 8-1 may help facilitate the conversion to a selected common unit.

Table 8-1
Conversion Factors from Various
Gamma Ray Log Units to API Gamma Ray Units

Logging Company	Conventional Units	API Units
Schlumberger	1 μg Ra-eq/ton	16.50
Lane Wells (now Dresser-Atlas)		
Series 400 (scintillation counter)	1 radiation unit (R.U.)	2.06
Series 300 (Geiger counter)	2.02 counts/minute	1.00
Series 200 (ionization chamber)	1 standard unit (S.U.) = 100 RU	206.00
PGAC (now Dresser-Atlas)		
Type F (Geiger counter)	1 Microroentgen per hour	14.00
Type T or Ta (scintillation)		15.00
McCullough (now Baroid)	1 Microroentgen per hour	10.40

References

1. Alekseyev, F. A., "Radiometric Method of Oil and Gas Exploration," *Nuclear Geophysics,* Moscow, 1959, pp. 3-26. Reviewed in *Geophysical Abstracts,* 183-525, (October, 1960, p. 607.

2. _____ et al., "Results of Radio-chemical Investigations of the Kyurov-Dag Oil Field," *Nuclear Geophysics,* Moscow, 1961, pp. 160-176. Reviewed in *Geophysical Abstracts,* 199-286, (August, 1963), p. 729.

3. _____, "Radiometric Method of Direct Exploration for Oil and Gas," *Acta Geophysica Sinica,* 9, No. 1, (1960), pp. 47-64. Reviewed in *Geophysical Abstracts,* 190-515, July-September, (1962), p. 435.

4. _____ and R. P. Gottikh, "Mechanism of Origin of Radiometric Anomalies Over Oil Deposits" Translated in *International Geologic Review,* 8, No. 10 (1966), pp. 1157-1171.

5. Alparone, N., "Subsurface Anomalous Radioactivity in the Environment of Oil Fields," Master's thesis, The University of Texas, July, 1963.

6. Avadisian, A., "Surface and Subsurface Radiation Anomalies in Wellman Oil Field," Master's thesis, The University of Texas, August, 1965.

7. Baker, E. G., "Origin and Migration of Oil," *Science,* 129, No. 3353, (April 3, 1959), p. 871.

8. _____, "Distribution of Hydrocarbons in Petroleum," *AAPG Bulletin,* 46, No. 1, (1962), p. 76.

9. Baijal, S., "Geochemical Survey of the Hilbig Oil Field, Bastrop County, Texas," Master's thesis, The University of Texas, August, 1962.

10. Bell, K. G., C. Goodman and W. L. Whitehead, "Radioactivity of Sedimentary Rocks and Associated Petroleum," *AAPG Bulletin,* 24, No. 9, (September, 1940), pp. 1529-1547.

11. Bordenave, M., A. Combaz and A. Giraud, "Influence de l'origine des matières organiques at de leur degré d' évolution sur les produits de pyrolyse du kérogène," *Third International Con-*

gress of Organic Geochemistry, London, September, 1966, pp. 26-27

12. Buckley, S. E. et al., "Dissolved Hydrocarbons in Subsurface Waters," *Habitat of Oil,* AAPG, 1958, pp. 850-882.

13. Case, L. C., "Subsurface Water Characteristics in Oklahoma and Kansas," *Problems of Petroleum Geology,* AAPG, 1934, pp. 855-868.

14. Dimitriyev, M. K. et al., "Experience in Application of Radiometric Investigations for the Direct Prospecting for Oil Deposits in the Bashkir ASSR," *Nuclear Geophysics,* Moscow, 1960, pp. 206-219. Reviewed in *Geophysical Abstracts,* 189-503, (April-June, 1962), p. 276.

15. Flerov, G. N. et al., "Use of the Methods of Atomic Physics in Oil and Gas Prospecting and Production," Proceedings Fifth World Petroleum Congress, New York, 1959.

16. Hodgson, G. W. et al., "Research Trends in Petroleum Genesis," *Eighth Commonwealth Mining and Metallurgical Congress,* Vol. 5, *Proceedings-Petroleum,* Melbourne, Australia, 1966, pp. 9-19.

17. _____, "The Water and Hydrocarbon Cycles in the Formation of Oil Accumulations," *Recent Researches in the Fields of Hydrosphere, Atmosphere and Nuclear Geochemistry,* Tokyo, Japan, 1964.

18. _____, "Alkanes in Aqueous Systems • I • Exploratory Investigations on the Accomodation of C_{20}-C_{33} n—Alkanes in Distilled Water and Occurence in Natural Water Systems," *American Oil Chemists' Society,* Vol. 43, No. 4, (n.d.), pp. 215-222.

19. Hunt, J. M. and G. W. Jamieson, *Habitat of Oil,* AAPG: New York, 1958, pp. 735-746.

20. Lundberg, H., "What Causes Low Radiation Intensities Over Oil Fields?," *The Oil and Gas Journal,* 54, No. 52, (1956), pp. 192-195.

21. Meinschein, W. G., "Origin of Petroleum," *AAPG Bulletin, 43, No. 5, (May, 1959), p. 925.*

22. Munn, M. J., "Hydraulic Theory," *Economic Geology,* 4, No. 6, (1909), pp. 509-528.

23. Pirson, S. J., "Ground Gas Survey Is Promising Tool," *The Oil Weekly,* 91, No. 5, (October 10, 1938), pp. 34-44.

24._____, "Critical Survey of Recent Developments in Geochemical Prospecting," *AAPG Bulletin,* 24, No. 8, (August, 1940), pp. 1464-1474.

25._____, N. Alparone and A. Avadisian, "Implications of Log Derived Radioactivity Anomalies Associated with Oil and Gas Fields," Paper V, SPWLA Symposium, Tulsa, Oklahoma, May, 1966, p. 25.

26._____, "Projective Well Log Interpretation," *World Oil,* 157, Nos. 5 and 6, (October, November, 1963), pp. 116-120 and pp. 88-92; and 159, Nos. 2–6 (August-November, 1964), pp. 68-72, 83-86, 180-182 and 156-166.

27. Shneyerson, V. B. et al., "Determination of the Surface Properties of Natural Samples of Rocks and Soil in Prospecting for Oil and Gas by the Radiometric Method," *Nuclear Geophysics,* Moscow, 1961, pp. 216-228. Reviewed in *Geophysical Abstracts,* 199–288, (August, 1963), p. 729.

28. Sikka, D. B., "Radiometric Survey, Redwater Oil Field, Alberta, Canada," Ph.D. dissertation, McGill University, Montreal, 1959.

29. Smith, P. V., Jr., "Studies on Origin of Petroleum Occurrence of Hydrocarbons in Recent Sediments," *AAPG Bulletin,* 38, No. 3, (March, 1954), pp. 337-404.

30. Stevens, N. P., E. E. Bray and E. D. Evans, "Hydrocarbons in Sediments of Gulf Coast of Mexico," *AAPG Bulletin,* 40, No. 5, (May, 1956), pp. 975-983.

31. Tilloy, R. and A. Monchaux (in collaboration with M. Ortynski, M. Dardene and M. Poupon), "Radioactivité et Recherches de Pétrole," Fifth Petroleum National Congress—Association Francaise des Techniciens du Pétrole (AFTP), Le Touquet, France, June 9-12, 1965.

32. Trask, P. D., "Deposition of Organic Matter in Recent Sediments," *Problems of Petroleum Geology,* AAPG, 1934, pp. 27-33.

33._____ and H. W. Patnode, *Source Beds of Petroleum,* AAPG 1942.

34. Weeks, L. G., "Habitat of Oil and Factors that Control It," *Habitat of Oil,* AAPG, 1958, p. 53.

35. Weller, J. M., "Compaction of Sediments," *AAPG Bulletin, 43, No. 2, (February, 1959), pp. 273-310.*

36. Whitmore, F. C., "Fundamental Research on Occurrence and Recovery of Petroleum," *American Petroleum Institute,* 1946-1947, p. 95.

37. Yermakov, V. I. et al., "Results of Investigation of the Natural Gamma Field in Oil Bearing Regions by the Methods of Airborne and Land Radiometric Survey," *Nuclear Geophysics,* Moscow, 1959, pp. 264-278. Reviewed in *Geophysical Abstracts,* 189-493, (April-June, 1962), p. 274.

38. ZoBell, C. E., "The Role of Bacteria in the Formation and Transformation of Petroleum Hydrocarbons," *Science,* 102, No. 2650, (1945), pp. 364-369.

Hydrogeology II:
Geostatic Equilibrium

Sedimentary rocks during burial maintain hydrostatic fluid pressure within their pore space if fluids within the sediments are allowed to escape as the sediments compact. If they are not permitted to escape, compaction is retarded; and the fluid pressure rises (sediments become overpressured) and ultimately approaches the pressure that over-lying rocks and contained fluids exert.

Standard bottom hole pressure measurements can determine the actual fluid pressure in a given permeable formation. Determining the fluid pressure in shales, with their low permeability, has previously been difficult or impossible. Chapter 9's discussion of the prediction techniques for abnormally high pressures follows Ham's,[15] Hottman's and Johnson's[16] work.

Density, acoustic and resistivity logs can determine the fluid pressure within the pore space of shales. The method establishes relationships between the shale transit time, shale density or shale resistivity and depth for normally pressured formations. In such formations with a plot of transit time versus depth or of shale density versus depth, a linear relationship generally appears; whereas on a plot of resistivity versus depth, a nonlinear trend exists.

Deviation of observed transit time, density or resistivity values from established normal compaction trends under hydrostatic pressure conditions measures pore fluid pressure in the shale and in adjacent sealed permeable formations. Actual pressure measurements in adjacent permeable formations has empirically established this. Using these data and this method predicts fluid pressure from acoustic, density and resistivity measurements with an accuracy of approximately 0.04 psi per foot, or about 400 psi 10,000 feet deep.

Knowing the first occurence of overpressure, and of the precise pressure-depth relationship in a geologic province, improves drilling techniques, casing programs, completion methods, hydraulic fracturing and reservoir evaluations. As much as one third of the costs of drilling deep wells on the Gulf Coast have been saved by unnecessary expenditures in casing and drilling mud and by increased drilling rate from using a minimum mud weight. Drilling safety also increased; and fewer mud circulations and wells were lost and fewer drill pipes were stuck.

Formation Pressures

Normal pressure at a given depth refers to formation pressure, which approximately equals the hydrostatic head of a water column of equal depth. If a normally pressured formation were opened to the atmosphere, a water column extending from reservoir to ground levels would balance the formation pressure. On the Gulf Coast, the shallow, predominantly sand formations contain fluids under hydrostatic pressure. These formations are normally perssured or have a normal pressure gradient. The normal pressure gradient on the Gulf Coast is approximately 0.465 psi per foot deep (Cannon and Craze[6]).

Formations with pressures higher than hydrostatic vary in depths in many geologic sequences, especially in thick Tertiary sediments. These formations are referred to as "abnormally pressured," "abnormally high pressured" or "overpressured." Jones[18] has observed formation pressured up to twice the hydrostatic pressure.

Compaction and Fluid Relationships

The cause of overpressured formations in Tertiary sections of the Gulf Coast as well as in many other Tertiary sedimentary basins primarily is the result of the compaction phenomenon. In the Gulf Coast area of South Louisiana, present sedimentation processes are essentially identical to those which occurred in Oligocene, Miocene and Pliocene eras. Mud and sand are brought into the Gulf of Mexico mainly by the Mississippi river and by other lesser streams.

Westerly offshore currents spread the suspended solids on the Gulf floor, creating sand and shale layers which—after burial under younger sands and shales—became the reservoirs and source rocks for the Gulf Coast oil and gas fields. Shales at shallow depths were originally soft muds containing a high percentage of sea water, possibly 80 percent or more. As new sediment layers added to this huge unconsolidated sequence of mud and ooze, the overburden weight squeezed water from the shales in a rather continuous, but exponentially decreasing rate.

This compaction pressed the thin platey mineral grains of clays against each other to get sufficient bearing strength from grain to grain contact to support the weight of the overlying rock. The water so squeezed out transported oil and gas to their primary loci of accumulation in porous sands. In this process the original muds lost more than half of their total bulk volume when buried approximately 10,000 feet deep. During this compaction, huge volumes of water excaped from the shales and were removed from the sedimentary section. This expelled water could only escape upward, back to the surface, to the margin of the sedimentary basin. The compaction involved transporting huge quantities of water from great depths in the sedimentary basin to near surface sediments, where it was discharged back to the sea. Oil and gas, probably as colloidal suspensions, were transported by this water and eventually trapped toward the shore. South Louisiana oil fields contain oil and gas transported from a deeper and more southerly point of origin to their present traps often many miles to the north.

Of particular interest are conditions in which the water squeezing and water removal are interrupted. Normal and gravity faults' developing or sand lenses' terminating most commonly cause this. In Gulf Coast sedimentation, loading by the weight of a continuous accretionary overburden has caused numerous down-to-the-coast gravity faults. These faults, as well as more complex faulting systems due to rock salt flowage, blocked the water flow, severing the escape channels. As the loading continued relentlessly, pore pressure continued to increase, but the fluids could no longer escape and were forced to support partially or totally the over-burden weight.

This process eventually may create a maximum pressure gradient of 1.0 psi per foot deep. When this is reached, the trapped fluids are supporting the entire weight of the sedimentary over-burden. Fortunately, this condition is rare. Where water can escape without restriction up dip and back to sea, the pore pressure gradient is approximately 0.465 psi per foot deep. Abnormal pressures in South Louisiana range from 0.465 to 1.0 psi/foot, depending on the rock load the contained fluids carry.

Electric Resistivity Effect

When water may freely escape and, thus, when pore pressures are normal, shale compaction is largely a function of depth and—to a much lesser degree—a function of time. Hence, the deeper the burial depth, the greater the compaction degree and the greater the shale density.

On the induction electric log of a well which encountered normal pressures, the conductivity curve in shale sections shows a decreasing value versus depth. If selected conductivity readings are made and plotted versus depth, the points fall essentially on a straight line. In this procedure one must be cautious and avoid making readings in silty or sandy shales. Slight variations from a straight line that may appear result from differences in the shale's mineral composition. Such variations may be in kaolinite, montmorillonite or illite content. The amplified normal resistivity curve may be used similarly, since resistivity is the reciprocal of conductivity. Induction log's conductivity curve is preferred because of its stability and the negligible effects of hole size and mud resistivity. In each case the maximum shale conductivity or minimum shale resisitivity is plotted.

When shales contain a certain amount of water which can no longer escape, their electrical conductivity becomes greater than what it would normally be for the burial depth. This water, which is somewhat altered sea water and generally of greater salinity, has high electrical conductivity. The measured shale conductivity is essentially proportional to salt water in its pore space. The trapped salt water is proportional to the shale pore pressure. Hence, the shale electric conductivity reflects the shale pore pressure. In most cases, adjacent sand will have an equal pressure. Observing a shale's electrical conductivity can, therefore, estimate fluid pressure in an adjacent sand body.

Numerous factors cause the measured conductivity of shales to vary. One of these is variation in formation temperature. Another important factor is the contained water's salinity. Although the original trapped water had the salinity of sea water (approximately 35,000 ppm NaCl), the chemical and physical reactions of salt with the clay minerals caused a higher chloride content in the first water expelled at low overburden pressures. During the later stages of compaction, the expelled water becomes quite fresh because of the shales' salt-sieving effect at high temperatures and pressures (Powers[32]). Another factor is the rock's age.

More factors, including variations in logging equipment exist. In practice these difficulties are avoided by plotting shale conductivities versus depth in the upper part of the hole to establish a "normal trend." This technique considers many of the variables mentioned. Once a normal trend is established, deviations from it indicate the degree of pressure abnormality. Making readings at shallow depths, first, establishes a "trend line" for normally pressured sediments.

Certain care is required for selecting pure shales versus silty shales. One must observe the spontaneous potential curve opposite the selected conductivity points and select only those levels which indicate the purest shale. This normally occurs when SP and resistivity curves approach the center of the log simultaneously.

This trend technique recognizes normal pressures and determines the approximate pore pressure of abnormally pressured sands. The most useful information is the minimum depth at which abnormal pressures occur. With practice, this can readily be seen by scanning the log from top to bottom, observing the conductivity variations of thick shales. This begins with values at shallow depths of approximately 2,000 millimhos (0.5 Ωm). With increasing depth, this value will gradually diminish. In South Louisiana wells 16,000 feet deep, the value may be 400-600 millimhos (2.5 to 1.6 Ωm) if pressures are normal. If abnormal, shale values will increase, usually exceeding 2,000 millimhos (0.5 Ωm); in some diapiric shales, values will reach 4,000 millimhos (0.25 Ωm). Changes in trend indicate the beginning of the transition zone in shales under abnormal pore pressures.

Shale Compaction Effects

Workers in soil mechanics have well established the consolidation theory of a water-saturated clay. The compaction concept is simulated by a model with perforated metal plates separated by metal springs and enclosed in a cylindrical tube of water. Figure 9-1a[14] schematically represents such a model (Terzaghi and Peck[36]). The springs simulate grain-to-grain contacts between clay particles, and the metal plates simulate the clay particles. Manometers record the fluid pressure. Pressuring the uppermost plate, the height of springs between the plates remains unchanged as long as no water escapes from the system. Thus, in the initial stage, the equal and opposite water pressure entirely supports the applied pressure.

Conditions for stage A in Figure 9-1a overpressure the system. As water is allowed to escape from the system (stage B), the plates move downward slightly (the system compacts), and the springs carry part of the applied load. As more water is allowed to escape, the springs carry a greater share of the load. Finally, when sufficient water has escaped for the springs to attain compaction equilibrium (stage C), the springs and water pressure, which is now simply hydrostatic, support the applied load.

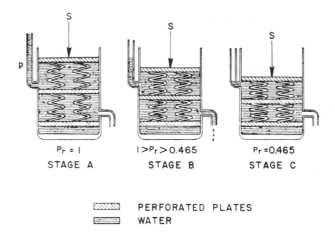

PERFORATED PLATES
WATER

Figure 9-1,a. Shale compaction model.

This model is analogous to a clay undergoing vertical compaction (Figure 9-1b[14]) in response to a vertical component of total stress—GS (overburden or geostatic pressure)—where

$$GS = 0.435 \, D_b \times h \qquad (9\text{-}1)$$

D_b is the mean value of the water-saturated bulk density of the overlying sediments, and h is the burial depth in feet. Rubey and Hubbert[35] have demonstrated the fluid pressure (P) and the grain-to-grain bearing stress (σ) of the clay particles support the GS load, where

$$\sigma = GS - P \qquad (9\text{-}2)$$

and σ is analogous to the support that springs in the Terzaghi-Peck model provided.

As Rubey and Hubbert have said, "The effective stress σ exerted by the porous clay (or by the springs in the model) depends solely upon the degree of compaction of the clay, with σ increasing continuously as compaction increases. A useful measure of the degree of compaction of a clay is its porosity ϕ, defined as the ratio of the pore volume to the total volume. Hence, we may infer that for a given clay there exists for each value of porosity ϕ some maximum value of effective compressive stress σ which the clay can support without further compaction."

Thus, the porosity (ϕ) at a given burial depth (h) depends on the fluid pressure (P). If the fluid pressure is abnormally high (greater than hydrostatic), porosity will be abnormally high for a given burial depth.

In Tertiary sediments of the Gulf Coast, thick shale intervals appear frequently. Many of these intervals are deep water marine shales containing isolated sands. These sediments essentially have been subjected only to vertical compaction due to the overburden's geostatic pressure.

In order for a shale to compact, fluids must be removed. Sands, which are highly permeable media, act as avenues of fluid escape. The scarcity of sands in thick shale sections reduces the rate of fluid removal from these shales. Fractures and nonsealing

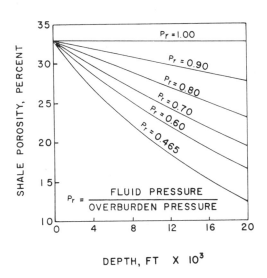

Figure 9-1,b. Porosity versus depth for average Gulf Coast shales. (Courtesy of D. G. Griffin, D. A. Bazer and SPE of AIME.)

faults can also act as avenues of fluid escape, but sands are the more important flow paths.

A useful measure of the compaction to which a clay has been subjected is its present porosity. Thus, an estimate of shale porosity as a function of depth should reveal the compaction degree. A section which is "undercompacted," with regard to its burial depth, is a section in which fluid pressure is abnormally large (in excess of hydrostatic pressure). The information various well logs record can infer the compaction degree. Methods by which this may be achieved follow.

Well Log Methods for Estimating Formation Pressures

Shale Acoustic Travel Time

Hottman and Johnson Method[16]

The acoustic log is a way to estimate the porosity of sedimentary rocks. Acoustic velocity of the compressional wave that various acoustic logs record is primarily a function of porosity and lithology. In shales the acoustic log essentially responds to porosity variations. Porosity's changing with depth can be so studied to grasp shale compaction. Investigating an acoustic log's response in normally pressured shales has indicated a straight line relation between the logarithm of shale travel time (Δt_{sh}) and depth.

An example of the relation for Miocene and Oligocene sediments is in Figure 9-2[16]. It illustrates the travel time decreases (velocity increases) with increasing burial depth, indicating that porosity decreases as a function of depth. This trend represents the "normal compaction trend" as a function of burial depth, and pore pressures exhibited within this normal trend are hydrostatic.

If overpressured formations appear, data points deviate from the normal compaction trend toward abnormally high transit times for a given depth, since the porosity is higher in undercompacted shales. Figure 9-3[16] illustrates such data. The deviation of a given point from the established normal compaction trend has

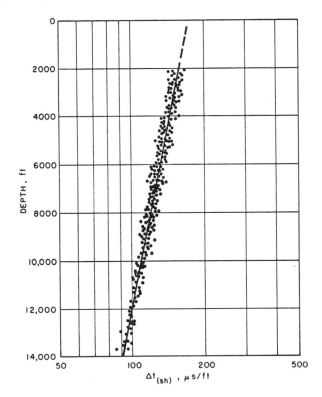

Figure 9-2. Shale traveltime versus burial depth for Miocene and Oligocene shales, upper Texas and Louisiana Gulf Coast. (Courtesy of C. E. Hottman, R. K. Johnson and SPE of AIME.)

been related to the observed pressure in adjacent reservoir formations. Figure 9-4[16] is a schematic plot of Δt_{sh} and pressure for Miocene and Oligocene formations. The average deviation from the line representing these data in Figure 9-5[16] is 0.04 psi per foot.

Estimating the formation pressure of reservoirs from acoustic log data in adjacent shale requires these steps:

1. Plotting Δt_{sh} versus depth on semilog graph paper (as in Figures 9-4 and 9-5) establishes the normal compaction trend for the particular area.

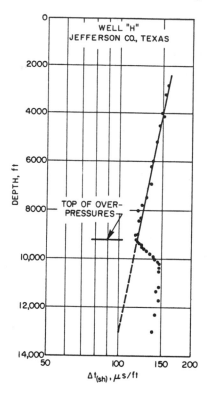

Figure 9-3. Shale traveltime versus burial depth. (Courtesy of C. E. Hottman, R. K. Johnson and SPE of AIME.)

2. A similar plot is made for the well in question at certain depths.

3. The top of the overpressured formations is found by noting the depth at which the plotted points deviate from the normal compaction trend line.

4. Following these steps locates the pressure of a reservoir at any depth:

(a) The deviation of Δt for adjacent shales from the extra-polated normal trend line is measured, as in Figure 9-4.

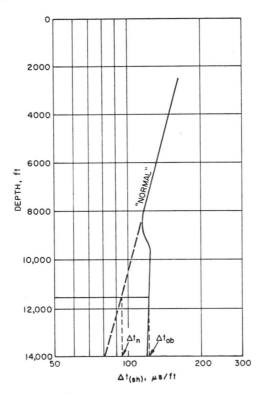

Figure 9-4. Schematic representation of shale traveltime versus burial depth. (Courtesy of C. E. Hottman, R. K. Johnson and SPE of AIME.)

(b) Figure 9-5 then finds the fluid pressure gradient (FPG) corresponding to the $\Delta t_{sh} - \Delta t_{n(sh)}$ value.

(c) The FPG value is multiplied by the depth to obtain the reservoir pressure.

Ham Method[15]

Ham has suggested a way to determine overpressured formations, which applies to the Louisiana Gulf Coast.

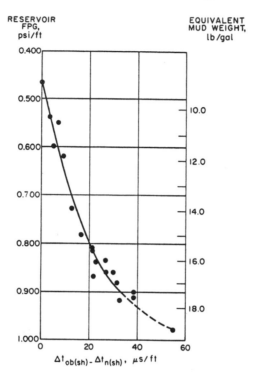

Figure 9-5. Relation between differential traveltime (Δtsh - Δtsh normal) and reservoir fluid pressure gradient. (Courtesy of C. E. Hottman, R. K. Johnson and SPE of AIME.)

The South Louisiana technique bases an empirically derived chart (Figure 9-6 [15]) on many observations of the transit time versus pressure relationship in the coastal area near New Orleans. The chart consists of a grid on which two families of curves are inscribed: dashed curves as contours of equal formation pressure and dotted curves as contours of equal mud weight. The solid curve bounding the chart below 6,000 feet on the right is the normal pressure gradient (usually 0.465 psi/per foot on the Gulf Coast). Above 6,000 feet the slope change reflects the effect on sonic transit time of undercompacted formations at shallow depths.

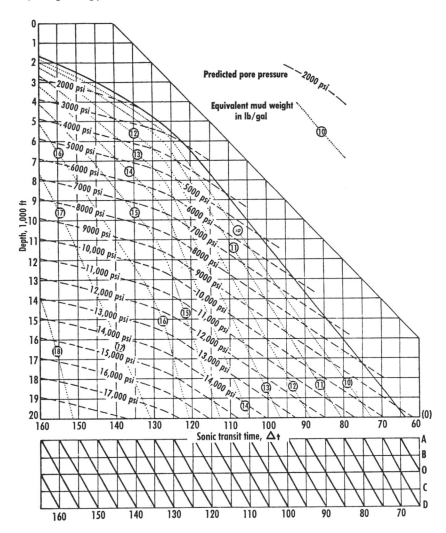

Figure 9-6. Sonic-depth pressure prediction chart. (Courtesy of H. H. Ham and *The Oil and Gas Journal*.)

Predicting formation pressure from this chart involves these steps:

1. Enter the chart in abscissa with the shale transit time and in ordinate with the corresponding depth.

2. The point so located will lie on or between dashed lines representing formation pressure and on or between dotted lines representing mud weight.

3. Interpolate between appropriate lines to determine the formation pressure and the mud weight required to balance this pressure.

The "normal" formation pressure at the particular depth is determined by multiplying the formation depth by 0.465. The difference between the normal and formation pressures is the amount by which the formation is abnormally pressured. For example, if one reads in a shale at 12,500 feet a travel time of 132 μsec/feet, these values entered on the chart give a 10,200 psi formation pressure. Thus, at this depth,

"Normal" pressure = 12,500 x 0.465 = 5820 psi.

So, the formation is overpressured accordingly:

Overpressure = 10,200 - 5820 = 4380 psi.

The chart shows that the mud weight required to balance the formation pressure is 15.8 pounds/gallon.

Shale Resistivity Method

Gulf Coast log analysts have observed and recognized that shale resistivity decreases in overpressured formations. This phenomenon has detected "sheath material" near salt domes in various areas and is considered a qualitative indication of high formation pressure gradients. The next logical step is to determine how shale resistivity can estimate actual formation pressures, considering that numerous factors affect shale resistivity. Many parameters which influence the resistivity of reservoir rocks also affect shale resistivities. Among these, the more important are *porosity, temperature, salinity of the contained fluid* and *mineral composition.*

Hottman and Johnson Method[16]

As in the acoustic method, this establishes a trend of shale resistivity versus depth for hydrostatically pressured shales in a

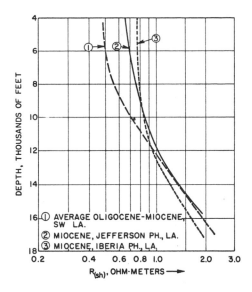

Figure 9-7. Shale resistivity versus burial depth. (Courtesy of H. H. Ham and SPE of AIME.)

given area. Typical trends of data from hydrostatic pressured shale sections appear in Figure 9-7[16] These data points are from standard electrical resistivity logs; the amplified short normal device is used because of its readability and because of negligible borehole corrections in the range of resistivities considered.

These trends reflect the normal compaction resistivity trend as a function of depth in a given area. When overpressured formations appear, shale resistivity data points deviate from the normal compaction trend toward lower resistivity values due to exceptionally high porosity. An example of resistivity depth plot is in Figure 9-9a.

Deviation of a given point from the established normal compaction trend has been related to the observed pressure gradient in adjacent reservoir formations. Figure 9-8[16] plots the pertinent information that establishes this empirical relation. Maximum deviation of the data from the smooth curve in Figure 9-8 is approximately 0.08 psi per foot, and the average deviation is

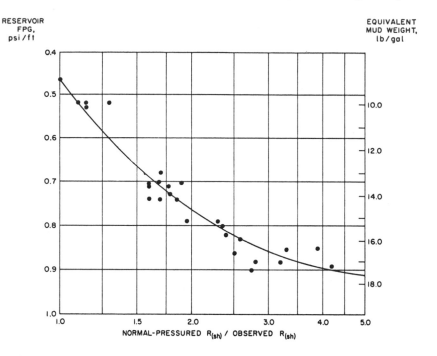

Figure 9-8. Relation between the shale resistivity ratio (Rsh normal/Rsh) and reservoir fluid pressure gradient. (Courtesy of C. E. Hottman, R. K. Johnson and SPE of AIME.)

approximately 0.04 psi per foot. The figure illustrates that an increase in the ratio of extrapolated normally pressured shale resistivity to actual recorded shale resistivity signifies an increase in formation pressure gradients. The trend in Figure 9-8 should be considered an example plot and used only as a guide until actual pressure and log data are obtained for the region under study. Estimating the formation pressure of reservoirs from adjacent shale resistivity data follows these steps:

1. The normal compaction trend for the particular area comes from plotting the shale resistivity from the amplified short normal device versus depth on semilogarithmic paper. Numerous wells in the area should be examined to establish an average and reliable normal resistivity compaction trend.

2. A similar plot is made for the well in question.

3. Noting the depth at which the plotted points deviate from the trend line locates the top of the overpressure formations.

4. These give the pressure gradient of a reservoir at any depth:

   (a) When the ratio of the extrapolated normal shale resistivity to the observed shale resistivity is determined;

   (b) When the fluid pressure gradient (FPG) corresponding to the calculated ratio is read from Figure 9-8 or from a similar calibration curve valid for the area under study.

5. Reservoir pressure is obtained by multiplying the FPG value by the depth (repeating this procedure at numerous depths can construct a pressure gradient profile for the well.)

Conditions of the borehole and of the adjacent disturbed formations influence the acoustic and the resistivity log recordings. Generally, employing normal borehole correction procedures can overcome these effects. If drilling causes a large temperature disturbance, one of the longer spaced resistivity devices may have to determine shale resistivity. The caliper survey should determine zones of extreme borehole enlargement in which erroneous shale transit times may be recorded from weak signals and cycle skipping.

In general, fresh or brackish-water zones at considerable depths may lead to anomalously high resistivity values and make it extremely difficult, if not impossible, to use the resistivity method for pressure estimation. Acoustic log data frequently can be used in such areas. Variations in shale clay mineralogy and nonclay constituents create difficulties with both techniques. Properly selecting data points can greatly reduce this problem. Zones of low SP deflection and uniform resistivity and/or sonic readings should always be selected.

Example of Application

Hottman's and Johnson's example of a well drilled in Cameron Parish, Louisiana, illustrates estimating formation pressures that they proposed from shale properties. This well penetrated several thousand feet of overpressured sediments; three actual bottom hole pressure measurements verify the accuracy of the methods.

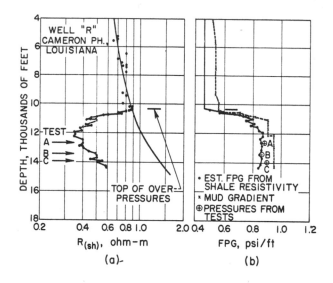

Figure 9-9,a and b. Example of estimation of abnormal formation pressure from resistivity log. (Courtesy of C. E. Hottman, R. K. Johnson and SPE of AIME.)

In Figure 9-9a,[16] shale resistivity is plotted against depth on semilog paper. It shows the average normal resistivity trend fitted to the previously discussed data. Shale resistivity points' departure from the normal trend can determine the top of the overpressured zone occurring at approximately 10,400 feet. The ratio of the observed resistivity to the normal trend resistivity at the same depth is determined at numerous levels. Figure 9-8 locates the fluid pressure gradient at each level. Figure 9-9b plots these data. For comparison, the mud column pressure gradient used while the well was being drilled is shown. Also, three bottom hole pressure readings were obtained from tests at 12,700, 13,500 and 13,900 feet. The calculated pressure gradients are within 0.04 psi per foot of the measured gradients.

Observed shale travel times in the same well are plotted against depth in Figure 9-10a.[16] A straight line is drawn through the shallow shale normal compaction trend. Observed points deviate again from the normal compaction trend at approximately 10,400

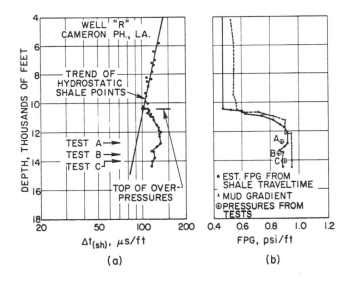

Figure 9-10, a and b. Example of estimation of abnormal formation pressure from acoustic log. (Courtesy of C. E. Hottman, R. K. Johnson and SPE of AIME.)

feet and confirm the top of the overpressured formation. Shale transit times' departure from the trend line is measured at numerous levels. Figure 9-5 gives the corresponding pressure gradient; Figure 9-10b plots the resultant trend of pressure gradient with depth. The mud column hydrostatic gradient and the measured pressure gradients from drillstem tests are compared. The estimated pressure gradients and the measured pressures correspond satisfactorily.

Ham's Method from Resistivity Logs[15]

Sonic log data generally provide reliable pressure estimations. However, sonic logs are not run in all wells or through all abnormally pressured zones. Thus, Ham makes a study parallel to the sonic log method, using shale resistivity as the pressure indicator. Figure 9-11[15] shows the results of cross plotting many shale resistivities against depth. Resistivity values were taken from the amplified short normal curve.

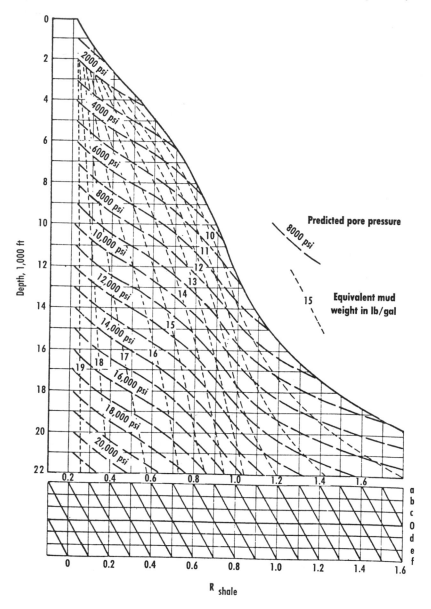

Figure 9-11. Chart for estimating abnormal formation pressure from shale resistivities in the Louisiana Gulf Coast. (Courtesy of H. H. Ham and *The Oil and Gas Journal.*)

Numerous variables other than compaction affect shale resistivity. The major ones are *salinity, shale mineral composition* and *formation temperature*. Possible salinity and matrix changes should be observed so that anomalous results may be recognized. The Figure 9-11 graph did not introduce any correction for these variables.

When shale resistivities are plotted versus depth, the plotted points indicate that the curve should show an increase in resistivity from 7,000 feet deep to the surface. This is due primarily to surface waters becoming somewhat saline with depth. It becomes necessary, therefore, to calculate the portion of the normal resistivity compaction curve above 7,000 feet.

Athy[1] estimates that clays and muds have lost 20 percent of their original volume at 1,000 feet; 35 percent at 2,000 feet; 40 percent at 3,000 feet; and 47 percent at 6,000 feet. Starting at the surface with a mud of 64 percent porosity, shale porosity was calculated using the volume reduction relationship that Athy has found. Using the porosities thus obtained, Ham computes hypothetical shale resistivities for shallow depths. In these computations, water resistivity was assumed constant at 0.06 ohm-m. Then, normal shale resistivities less than 6,000 feet deep were computed by the following equation:

$$R_{shale} = 0.048 \ \phi^{-2.15} \qquad\qquad (9\text{-}3)$$

With shale resistivity versus the normal pressure gradient, the Figure 9-11 chart was constructed like Figure 9-6. Again, a grid appears below the chart to shift the resistivity scale to compensate for local variations. In such a shift, the normal slope assumably remains unchanged.

Pressure determinations using the Figure 9-11 resistivity chart have been satisfactory despite many variables in the method. Readings must be made in clean shales. The SP or gamma ray curves should be at, or near, the shale base line in the selected reading intervals. Ham shows that limitation of the resistivity method is that it does not yield satisfactory results in abnormally pressured zones above 6,000 feet due to variations in R_w.

Shale Density Method

Shale density varies, of course, with the compaction degree and with the remaining fluid-filled porosity. This expresses shale density (D_{sh}):

$$D_{sh} = \phi \times D_w + (1 - \phi) D_g \qquad (9\text{-}4)$$

where

D_g = grain density (Generally, an average value of 2.65 satisfies most clay particles in shales.)

D_w = water density that fills pore space (D_w = 1.05 is a good average value for the salinity ranges encountered.)

ϕ = shale porosity, which varies from 10 percent under high degrees of compaction to 90 percent in the original state of mud and ooze.

The shale bulk density variation is thus from 1.2 to 2.4 grm/cm^3. Shale density may be determined for consolidated shales accordingly: (1) while drilling, from shale cuttings that did not disintegrate during drilling; (2) after drilling, from a density log.

While drilling, two techniques commonly measure cutting shale density.

1. *A calibrated fluid column:* Rogers[34] has described the technique. It uses two completely miscible liquids which, when standing in a vertical cylinder, establish a linear density gradient that glass beads of known density easily calibrated. Such beads, when dropped in the cylinder, rest and float at a level equal to their individual densities. Similarly, shale particles will float at levels equal in density.

2. *Volume measurement of shale cuttings of known weight*: A mercury pump makes such measurements. The technique is a standard core analysis procedure that Boatman[2] describes.

Measuring shale density from drill cuttings and studying their variation with depth as a way to study the proximity to high pressure formations have an advantage over the logging technique, since results may be obtained before drilling trouble possibly

Density - Shale Cuttings - GM/CC ➔

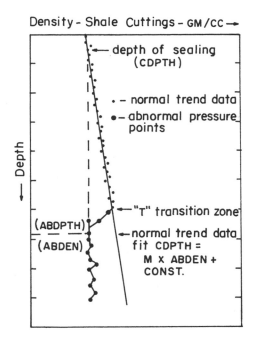

Figure 9-12. Shale cuttings density versus burial depth (Gulf Coast area). (Courtesy of D. G. Griffin, D. A. Bazer and the SPE of AIME.)

develops. After drilling close to a suspected high pressure formation, a density log may be run (preferably of the borehole compensated type) and a direct reading of shale density is made. Density gradients are observed directly on the recorded log.

Depth of Sealing Method

Griffin and Bazer[14] have proposed several ways to calculate formation pressures from shale density measurements, the simplest one being the depth of sealing method. *Depth of sealing* (DS) is that depth at which normally compacted shales would have the same density as the undercompacted shale. Figure 9-12 shows a graphic way to determine DS. It simply involves vertically projecting the undercompacted shale density to the sloping normal density versus depth.

The shale's pore pressure (FP) is then obtained by
$$FP = h - 0.535 \ DS \qquad (9\text{-}5)$$
where

h   = depth in feet;
DS= sealing depth in feet.

Example of Application

An example density log is in Figure 9-13.[14] The normal density-compaction gradient above 12,500 feet indicates that the depth of sealing would be at 7,000 feet. The abnormal formation pressure at 13,500 feet is then obtained by:

FP = 13,500 - 0.535 x 7,000 = 9655 psi.

Ham's Method from Density Logs[15]

Ham's chart in Figure 9-14[15] estimates pore pressure by shale density measurements. It is constructed like the charts for sonic and resistivity data, Figures 9-6 and 9-11. However, it incorporates data from fewer wells. Formation pressure and mud weight are estimated by cross plotting bulk shale density and depth and interpolating that point between lines of equal pressures and mud weight.

Theoretically, density measurements should detect porosity changes in shales accurately because relatively few variables enter the relationships between porosity and formation density. However, in practice, precisely measuring density in shale formations is hampered by rugosity of the borehole wall and by shale alteration near the borehole. Furthermore, the limited data that led to the Figure 9-14 chart indicate less resolution between normal and abnormal pressures than the sonic pressure chart shows. Presently, sonic and resistivity data place more confidence in pressure estimations. Increased experience with density derived pressures should lead to greater confidence in the method.

Geologic Significance of Abnormal Pressures

The discovery of abnormally high pressures in thick sedimentary sections has many theoretical and practical implications

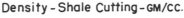

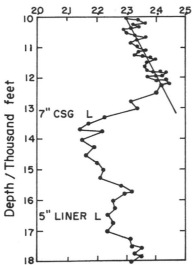

Figure 9-13. Shale cutting density versus burial depth above and within an abnormal pressure zone. (Courtesy of D. G. Griffin, D. A. Bazer and SPE of AIME.)

relative to the behavior of connate water during geologic time. The following observations are significant:

1. Water salinity 8,000 to 10,000 feet deep is rather variable and relatively unpredictable; but, below those depths, a pressure seal or geostatic barrier exist and the shaly formations are under-compacted (i.e., they have higher porosities than expected).

Pore pressure is higher than hydrostatic, the overpressure presumably being due to pore water's ability to leak at a restricted rate. If the pore water is completely confined and nonmovable, the pore pressure attains geostatic pressure of 1.0 psi/feet deep.

2. Totally or partially reducing vertical water escape during compaction, the heat's rate of escape also greatly reduces. Water movement is the greatest carrier of heat from the earth; thus, overpressured rocks become overheated and store geothermal energy. They act as thermal sinks and may be considered as some of the largest reserves of geothermal energy for the future. The geothermal gradient changes abruptly in slope when entering the

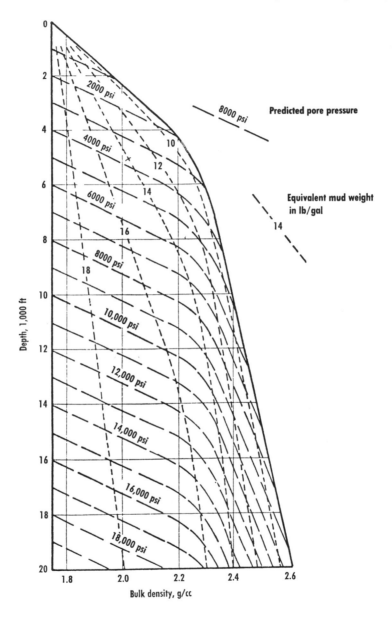

Figure 9-14. Chart for estimating abnormal formation pressure from shale density (from log or cuttings). (Courtesy of H. H. Ham and *The Oil and Gas Journal.*)

undercompacted shales. In the transition zone from normal to abnormal compaction, the geothermal gradient may become as high as 6°F per 100 feet deep (Jones[18]).

3. Shales act as salt or ion sieves, especially at high temperatures; water expressed out of shales under compaction is nearly fresh. As they are nearly perfect semipermeable membranes, it is thus natural to expect

$$\text{Shale porosity x Pore water salinity} = \text{Constant} \qquad (9\text{-}6)$$

a remarkable relation which Overton and Timko[31] have satisfactorily verified. These authors have suggested this relation for solving many practical geological problems such as these:

(a) Detecting normally compacted sediments when salinity increase with depth exponentially;

(b) Detecting overcompaction and abnormally low formation pressures by observing supernormal formation salinity;

(c) Detecting undercompaction and abnormally high formation pressures by observing subnormal formation salinity;

(d) Detecting stress relief from faulting by observing a gradually changing subnormal salinity;

(e) Detecting an extensive permeable sand by a salinity trend rapidly returning to normal with depth in a high pressure environment;

(f) Detecting proximity to abnormally high pressures ahead of the drill by observing salinity deviations with depth from normal trend.

4. Shales are nearly perfect semipermeable membranes across which osmotic pressures attain 95 percent of the theoretical calculated values. A salinity difference of 1N NaCl across a shale partition yields 690 psi differential osmotic pressure. With higher contrasts, tremendous pressure differences beyond the strength of the shale barriers are possible. Thus, such osmotic pressure differences may initiate incipient faults and diapir movements. High pressure waters act as a tectonic lubricant in thrust faulting (Rubey and Hubbert[35]).

5. Extreme pressures in shales release their pore water in three steps.

(a) Compaction and pore volume reduction expresses out pore water until clays hold only two layers of water molecules.

(b) Heating under great burial depth and a heat sink formation remove another monolayer of water molecules. By now porosity is 10 to 20 percent.

(c) Final removal of the last monolayer of molecular water results from diagenesis of montmorillonite into illite by potassium ions (Jones[18]).

In the latter stages of compaction and water removal from shale, water expressed out into adjoining sands is very fresh. This explains the fresh water aquifers at great depths, particularly in South Texas. Fresh water that clay dehydration releases may be as much as half the volume of montmorillonite in such clay (Jones[18]).

6. Expulsion of shale water into adjacent sands and aquifers causes unexpected reservoir engineering performances, such as producing large volumes of hydrocarbons without appreciable pressure drop even when no water encroachment per se exists. The water influx is from adjacent shales and is fresh. Wallace[41] reports some examples where water produced from deep high pressure distillate fields became fresher with time and without appreciable pressure drop.

Applications of Overpressure Detection

Practical uses of overpressure detection are obvious and of great economic importance.

1. Locating the high pressure transition zone and knowing where to expect the impending high pressure zone while drilling is one of the most important, necessary items to effect the minimum drilling cost:

(a) By predicting the mud weight required to overcome formation pressure without fracturing and facing loss circulation problems and well blowout;

(b) By determining casing seats and protecting low pressure formation from high pressures;

(c) By selecting logging depths;

(d) By determining fracturing gradient (FG). In hard rocks, the formula $FG = 1/3 (1 + 2 PP/h)$ is satisfactory; "PP" is pore pressure and "h," the depth, whereas in soft rocks (Gulf Coast) it is necessary to introduce variations of the Poisson ratio versus depth.

2. It manages fresh water supplies and prevents their contamination by high pressure salt water flow.

3. It determines reservoir limits and compartmentalizes independent fault blocks separated from each other by osmotic barriers.

4. It explains abnormal reservoir behavior in formations that are presumably sealed and volumetric.

References

1. Athy, L. F., "Density, Porosity, and Compaction of Sedimentary Rocks," *AAPG Bulletin,* 14, No. 1, (1930), pp. 1-35.

2. Boatman, W. A., "Measuring and Using Shale Density to Aid in Drilling Wells in High-pressure Areas," *Journal of Petroleum Technology,* 19, No. 11, (1967), pp. 1423-1429.

3. Bogomolov, Y. G., "Geotemperature Regime," *International Association Science Hydrology Bulletin,* No. 4, (1967), pp. 86-91.

4. Borel, W. J. and R. L. Lewis, "Ways to Detect Abnormal Formation Pressures," Part I, "Geopressure Predictions by Log Analysis," *Petroleum Engineer,* 41, No. 8, (July, 1969), pp. 49-63.

5. Burst, J. F., "Diagenesis of Gulf Coast Clayey Sediments and Its Possible Relation to Petroleum Migration," *AAPG Bulletin,* 50, 3, (1966), p. 607.

6. Cannon, G. E. and R. L. Craze, "Excessive Pressure and Pressure Variations with Depth of Petroleum Reservoirs in the Gulf Coast Region of Texas and Louisiana," *Transactions AIME,* 127, (1938), p. 31.

7. DeSitter, L. U., "Diagenesis of Oil-field Brines," *AAPG Bulletin,* 31, No. 11, (1947), pp. 2030-2040.

8. Dickey, P. A., C. R. Shriram and W. R. Paine, "Abnormal Pressures in Deep Wells in Southwestern Louisiana," *Science,* 160, No. 3828, (1968), pp. 609-615.

9. Dickinson, George, "Reservoir Pressures in Gulf Coast Louisiana," *AAPG Bulletin, 37,* No. 2, (1953), pp. 410-432.

10. Elder, J. W., "Physical Processes in Geothermal Areas, In Terrestrial Heat Flow," *American Geophysical Union Geophysical Monograph 8,* 1965, pp. 211-239.

11. Foster, J. B. and H. E. Whalen, "Estimation of Formation Pressures from Electrical Surveys, Offshore Louisiana," *Journal of Petroleum Technology,* 18, No. 2, (1966), pp. 165-171.

12. Fowler, W. A., Jr., "Pressures, Hydrocarbon Accumulation and Salinities, Chocolate Bayou Field, Brazoria County, Texas," Paper 2226, Society of Petroleum Engineers, Houston, 1968.

13. Frederick, W. S., "Planning a Must in Abnormally Pressured Areas," *World Oil,* 164, No. 4, (March, 1967), pp. 73-78.

14. Griffin, D. G. and D. A. Bazer, "A Comparison of Methods for Calculating Pore Pressures and Fracture Gradients from Shale Density Measurements Using the Computer," Paper 2166, Society of Petroleum Engineers, Houston, 1968.

15. Ham, H. H., "New Charts Help Estimate Formation Pressures," *The Oil and Gas Journal,* 64, No. 51, (December 19, 1966), pp. 58-63.

16. Hottman, C. E. and R. K. Johnson, "Estimation of Formation Pressures from Log-derived Shale Properties," *Journal of Petroleum Technology,* 17, No. 6, (1965), pp. 717-722.

17. Jones, P. H., "Hydrology of Neogene Deposits in the Northern Gulf of Mexico Basin," Proceedings of First Annual Symposium on Abnormal Subsurface Fluid Pressures, Baton Rouge, 1967.

18._____, "Hydrodynamics of Geopressure in the Northern Gulf of Mexico Basin," Paper No. 2207, Society of Petroleum Engineers, Houston, 1968.

19. Kerr, P. E. and J. Barrington, "Clays of Deep Shale Zone, Caillou Island, Louisiana," *AAPG Bulletin,* 45, No. 10, (1961), pp. 1697-1712.

20. Kryukov, P. A., A. A. Zhuchkova and E. V. Rengarten, "Change in the Composition of Solutions Pressed from Clays and

Ion Exchange Resins," *Doklady Akademii Nauk USSR,* 144, No. 6, (1962), pp. 1363-1365.

21. Lee, W. H. K. and Seiya Uyeda, "Review of Heat Flow Data, in Terrestrial Heat Flow," *American Geophysical Union Geophysical Monograph 8,* 1965, pp. 87-190.

22. Matthews, W. R. and J. Kelly, "How to Predict Formation Pressure and Fracture Gradient from Electric and Sonic Logs," *The Oil and Gas Journal,* 65, No. 8, (February 20, 1967), pp. 92-106.

23. McKelvey, J. G. and I. H. Milne, "Flow of Salt Solutions through Compacted Clay," in *Clays and Clay Minerals,* New York: Pergamon Press, 1962, pp. 248-259.

24. McNitt, J. R., "Review of Geothermal Resources, in Terrestrial Heat Flow," *American Geophysical Union Geophysical Monograph 8,* 1965, pp. 240-266.

25. Meyerhoff, A. A. and B. W. Beebe (eds.), "Geology of Natural Gas in South Louisiana," in *Natural Gases of North America,* AAPG Memoir 1968.

26. Meyers, R. L. and D. C. Van Siclen, "Dynamic Phenomena of Sediment Compaction in Matagorda County, Tex.," Society Transactions, *Gulf Coast Association Geologist,* 14, (1964), pp. 241-252.

27. Moses, P. L., "Geothermal Gradients," *World Oil,* 152, No. 6, (1961), pp. 79-82.

28. Mullins, John D., "Some Problems of Superhigh-Pressure Gas Reservoirs in the Gulf Coast Area," *Journal of Petroleum Technology,* 225, (September, 1962), pp. 935-938.

29. Nevin, C. M. and R. E. Sherrill, "Studies in Differential Compaction," *AAPG Bulletin,* 13, No. 10, (January, 1929), pp. 1-37.

30. Ocamb, R. D., "Growth Faults of South Louisiana," *Society Transactions, Gulf Coast Association Geologist,* 11, (1961), pp. 139-175.

31. Overton, H. L. and D. J. Timko, "The Salinity Principle— A Tectonic Stress Indicator in Marine Sands," *The Log Analyst,* 10, No. 3, (May, 1969), pp. 34-43.

32. Powers, M. C., "Fluid Release Mechanisms in Compacting Marine Mudrocks and Their Importance in Oil Exploration," *AAPG Bulletin*, 51, No. 7, (1967), pp. 1240-1254.

33. Rochon, R. W., "Relationship of Mineral Composition of Shales to Density," *Society Transactions, Gulf Coast Association Geologist*, 17, (1967), pp. 135-142.

34. Rogers, Les., "Shale-density Logs Help Detect Overpressure," *The Oil and Gas Journal*, 64, No. 37, (September 12, 1966), pp. 126-130.

35. Rubey, W. W. and M. K. Hubbert, "Role of Fluid Pressure in Mechanics of Overthrust Faulting," Part II, "Overthrust Belt in Geosynclinal Area of Western Wyoming in Light of Fluid Pressure Hypothesis," *Geological Society America Bulletin*, 70, No. 2, (1959), pp. 167-206.

36. Terzaghi, K. and R. B. Peck, *Soil Mechanics in Engineering Practice*, New York: John Wiley and Sons, Inc., 1948.

37. Thomeer, J. H. M. A. and J. A. Bottema, "Increasing Occurrences of Abnormally High Reservoir Pressures in Boreholes and Drilling Problems Resulting Therefrom," *AAPG Bulletin*, 45, No. 10, (1961), pp. 1721 and 1730.

38. Thorsen, C. E., "Age of Growth Faulting in Southeast Louisiana," *Society Transactions, Gulf Coast Association Geologist*, 13, (1963), pp. 103-110.

39. Von Engelhardt, Walf and K. H. Gaida, "Concentration Changes of Pore Solutions during the Compaction of Clay Sediments," *Journal of Sedimentary Petrology*, 33, No. 4, (1963), pp. 919-930.

40. Wallace, W. E., "Will Induction Log Yield Pressure Data?," *The Oil and Gas Journal*, 62, No. 37, (September 14, 1964), pp. 124-126.

41._____ , "Water Production from Abnormally Pressured Gas Reservoirs in South Louisiana," Part I, *Society Transactions, Gulf Coast Association Geologist*, 12, (1962), pp. 187-193; Part II, Paper 2225, Society of Petroleum Engineers, Houston, October, 1968.

42. Weller, J. M., "Compaction of Sediments," *AAPG Bulletin*, 43, No. 2, (1959), pp. 273-310.

43. White, D. E., "Saline Waters of Sedimentary Rocks," in *Fluids in Subsurface Environments,* AAPG Memoir 4, pp. 342-366.

44. Williams, D. G., "Six Tools Help You in Selection of Protective Casing Seats," *The Oil and Gas Journal,* 64, No. 41, (October 10, 1966), pp. 149-160.

45. Young, Allen and P. F. Low, "Osmosis in Argillaceous Rocks," *AAPG Bulletin,* 49, No. 7, (1965), pp. 1004-1007.

46. Zierfuss, H. and G. Van Der Vliet, "Laboratory Measurements of Heat Conductivity of Sedimentary Rocks," *AAPG Bulletin,* 40, No. 10, (1956), pp. 2475-2488.

10

Hydrogeology III: Hydrodynamics of Infiltration

On land, reservoir rocks may outcrop where they are subjected to infiltration by meteoric waters. Such waters which travel down-dip under gravity and its resulting hydrostatic pressure may move vertically downward and laterally for considerable distance. This depends on the flow potential gradient and, more importantly, on their ability to be discharged somewhere from the aquifers. Such waters at the infiltration points are very fresh; but, as they advance down-dip, they dissolve salts from rocks and mix with connate waters. When they encounter a hydrocarbon accumulation underground, they may partially flush such accumulation, leading to tilted water-hydrocarbon interface or to pressure build-up by hydro-osmosis.

Water Driving Forces

The fallacies of many theories on oil origin, migration, accumulation and entrapment stem mainly from the rash assumption that fluid flow in the earth can be represented only by one law of fluid motion—namely, Darcy's equation. On the contrary, one must view the earth fluids in a vast reservoir, which contains a multitude of all possible chemical elements and compounds that are in different phases and solutions (solid, liquid and gaseous) eventually subjected to numerous forces.

Chapter 10 considers only those phenomena of fluid transfer that occur without nuclear, atomic and chemical composition changes and, additionally, pertains only to those fluid transport phenomena in sediments.

The earth fluids may never be considered as being at rest or in equilibrium at any particular time. This would require that all space derivatives of the driving potentials (to be discussed later) be zero everywhere within the sedimentary basins. While this is not impossible, it is very unlikely even at a single point. The driving potentials would tend to reach equilibrium over geologic time if no further disturbances occurred. However, continental erosion, sedimentation, mountain building and storms and waves constantly disturb the trend toward equilibrium.

Various compartments may appear in the earth reservoir (a) on a large scale (sedimentary basins, oceans, seas, etc.) or on a smaller scale (individual sedimentary layers, lenses, faults blocks, etc.); (b) separated by semipermeable membranes of a chemical nature (shales, marls, clays, etc.) or of a physical nature (capillary barriers) or separated by insulators to electrical conduction (salt masses) and perhaps to heat conduction, although a perfect vacuum is the only perfect heat insulator and vacuum does not exist on earth.

The earth reservoir simmered over geologic time and generated numerous products, many of which are economically desirable. (This chapter, however, deals only with oil and gas and their associated formation waters.) During geologic time, the sedimentary fluids were subjected to different impelling forces—*osmosis* (from the Greek word meaning *impulse*). These forces may be derived as space derivative or rate of spacial change of various potentials; the most common one is the "hydrostatic flow potential" (H) exerted on the fluid—$H = P + gDh$—where P is pressure, g is the acceleration of gravity, D is fluid density and h is height relative to a datum plane.

Darcy's equation gives the fluid flow derived from non-equilibrium in flow potential, $Q = -A\frac{K}{\mu}\frac{\partial}{\partial r}(P + g\,Dh)$, representing *hydro-osmosis*—a terminology more adequate than the common term *hydro-dynamics*. Its impelling force is the space derivative of the hydrostatic potential—namely, $\frac{\partial}{\partial r}$ (P + g Dh), r being any direction in space. In addition, earth fluids are subjected to impelling forces derived from electric potentials associated in the earth with telluric currents. Large scale fluid migrations are undoubtedly due to such electric currents which, in the earth's various hemispheres exhibit certain predominant directions. The resulting fluid flows are *electro-osmotic* effects. They are the reverse effect of streaming potential, or electro-kinetic effect.

Earth temperature, or thermal potential, highly varies with depth; the rate of temperature change, or geothermal gradient, varies similarly within sedimentary basins. It causes an impelling *thermo-osmosis* force chiefly directed vertically upward which explains the higher formation water salinities at greater depths in certain formations. (The Arbuckle limestone, a well-known case, contains water of very low salinities near its outcrops in Northeast Kansas and gradually becomes a saturated brine in Central Oklahoma).

The most neglected aspect of fluid flow in the earth is, perhaps, *chemi-osmosis*, or *osmosis*. This phenomenon is responsible for most living cell processes and requires the working of a semipermeable membrane (i.e., of a partition permeable only to the liquid phase and not to the soluble salts or compounds). Such membranes abound in the earth as extremely fine-grained sediments, such as clays, shales, marls and chalks. They are permeable to water but not to some of the ions and to other chemicals in solution in formation waters. Fresh water tends to dilute highly concentrated connate waters through osmosis which thereby acquire abnormally high pressures for their depths when hydro-osmosis is impossible, i.e., when water escape to formations at lower flow potentials is impossible. The potential function whose space derivative leads to chemi-osmosis is this well-known chemical potential

$$\mu = \mu^o + RT\sum_{i=1}^{n} \ln x_i \qquad (10\text{-}1)$$

where:

μ° = standard potential that depends only on temperature and pressure

R = gas constant = 8.315 x 10^7 ergs per degree per mole

T = temperature (°K)

x_i = mole fraction of any compound in solution.

   The generalized *osmo-dynamic potential* (Φ) responsible for all four types of fluid transfer in the earth thus is

$$\Phi = \mu^{\circ} (T, H) + E + RT \sum_{i=1}^{n} \ln x_i \qquad (10\text{-}2)$$

and a generalized *osmotic force* or impulse may be derived from it by taking the space derivative of all the terms:

$$\frac{\partial \Phi}{\partial r} = \left(\frac{\partial \mu^{\circ}}{\partial r}\right)_T + \left(\frac{\partial \mu^{\circ}}{\partial r}\right)_H + \frac{\partial E}{\partial r} + RT \frac{\partial}{\partial r} \left(\sum_{i=1}^{n} \ln x_i \right) \qquad (10\text{-}3)$$

where:

   r = any direction in space

$\left(\dfrac{\partial \mu^{\circ}}{\partial r}\right)_T$ = hydro-osmosis (hydrodynamics) responsible for Darcy flow

$\left(\dfrac{\partial \mu^{\circ}}{\partial r}\right)_H$ = thermo-osmosis

$\dfrac{\partial E}{\partial r}$ = electro-osmosis

$RT \dfrac{\partial}{\partial r} \left(\sum_{i=1}^{n} \ln x_i \right)$ = chemi-osmosis

   The volume flux from each specific force may be calculated by multiplying each force by the cross-sectional area and by the appropriate retardation coefficient pertaining to each type flux.

Well Logs as Hydrodynamic Potential Measuring Devices

A properly conceived well logging program measures rocks' petrophysical properties that will control osmotic flow and measures the required osmotic potentials so fluid flow in the earth may be established.

The hydrostatic potential (H) is derived from measuring a bottom hole pressure device's elevation in the well and its representative pressure measurements. These are from properly conducted Formation Tester and Drill Stem tests. Most well logs and temperature logs measure the thermal potential (T°F). Unfortunately, such logs are seldom run under equilibrium conditions, and the results are all but useless for thermo-osmosis deductions.

The SP curve would measure the electric potential (E) if the latter were referred to a base value. Unfortunately, this is not and will not be done until it is practical to tie in all electric logs as a grid system.

The SP curve also measures the chemical potential (μ), as the former measures the molar salt concentration in formation waters. Accordingly, properly using the SP curve yields deductions on how chemi-osmosis is presently affecting the motion of formation waters.

Delineating Hydrodynamic Traps

The theoretical aspect of the hydrodynamic oil and gas entrapment in the earth is satisfactory when formation water is moving underneath a static accumulation of oil and/or gas and when oil and water differ in density; then, the oil-water interface is tilted. More specifically, we should consider the surfaces of constant capillary pressure as being tilted.

The degree of tilt depends on many factors: rate of water motion, differential viscosity and density of the fluids, absolute and relative permeabilities to the fluids, etcetera. The formation water must be "presently" moving for the water table to be tilted at the present time. The water's having been in motion in the past, or even in the recent geologic past, still doesn't warrant present day tilting.

A trap for oil and/or gas is an underground region of low flow potential in one or both of these fluids. The area is surrounded completely by closed equipotential surfaces and regions of higher flow potential or surrounded jointly by higher equipotential surfaces and an impermeable boundary (i.e., shaling-out, pinching-out, faulting and truncation). The technique presented here follows M. K. Hubert's presentation. [4,5,6]

According to these premises, the search for oil and gas traps turns into a search for underground regions of minimum flow potential for oil and/or gas. This involves the structural configuration of the rocks themselves, as geological and geophysical techniques determined, plus mapping the family of equipotential surfaces for oil and/or gas.

The flow potential of a particular fluid (water, oil or gas) may be represented as the hydrostatic head of that fluid (h_w, h_o or h_g), which is the height to which each particular fluid would rise in an open-top manometer filled with it—i.e., of fluid density corresponding to that of water (D_w), oil (D_o) or gas (D_g) expressed under average reservoir conditions of pressure, temperature and composition to which the manometer is connected. This is, of course, physically impossible to achieve; but, if the reservoir water pressure (P) is measured at a known depth and if it is referred as a height (Z_g)—g is gauge—above a reference datum level, fluid flow potentials in their respective phase may be written accordingly:

$$h_w = Z_g + \frac{P}{gD_w}$$

$$h_o = Z_g + \frac{P}{gD_o} \tag{10-3}$$

$$h_g = Z_g + \frac{P}{gD_g}$$

The above equations may readily be combined into the following by eliminating P:

$$h_o = \frac{D_w}{D_o} h_w - \frac{D_w - D_o}{D_o} Z_g$$

$$h_g = \frac{D_w}{D_g} h_w - \frac{D_w - D_g}{D_g} Z_g \tag{10-4}$$

Since the level at which pressure was measured is no longer material, Z_g may be replaced by the top of the structure's (Z_s) elevation in a particular well; h_o and h_g do not change in value because they are now independent of the depth at which pressure was measured.

For convenience, the above equations are multiplied by an amplification factor ($\frac{D_o}{D_w - D_o}$ or $\frac{D_g}{D_w - D_g}$, respectively). Then:

$$\frac{D_o}{D_w - D_o}\, h_o = \frac{D_w}{\rho_w - \rho_o}\, h_w - Z_s$$

$$\frac{D_g}{D_w - D_g}\, h_g = \frac{D_w}{D_w - D_g}\, h_w - Z_s \tag{10-5}$$

Functions $\frac{D_w}{D_w - D_o} h_w$ and $\frac{D_w}{D_w - D_g} h_w$ are the *amplified water flow potentials*—respectively, V_o (oil) and V_g (gas), which are both expressed in the same unit as Z (feet, meters, etc.). Functions V_o and V_g study the equilibrium conditions of oil and gas accumulations.

The amplified oil and gas flow potentials—respectively, U_o and U_g—are obtained by

$$U_o = \frac{D_o}{D_w - D_o}\, h_o = V_o - Z_s$$

$$\tag{10-6}$$

$$U_g = \frac{D_g}{D_w - D_g}\, h_g = V_g - Z_s$$

The U_o and U_g values may thus be readily obtained by subtracting from V_o or V_g, respectively, the structural elevation (Z_s) above a reference datum at every point on a map. Drawing maps of V_o, Z_s and U_o on transparent films most readily accomplishes this so that U_o may be obtained by the V_o-Z_s difference at each of the intersecting V_o and Z_s contour lines.

The U_o map is significant for oil accumulation under hydro-dynamic control combined with structural and/or stratigraphic control, whereas U_g controls gas accumulation under hydro-dynamic conditions. Closed contours on the U_o map are likely to delineate probable bubbles of oil accumulation in static equilibrium above an underlying drifting water. Studying gas accumulation under hydrodynamic conditions utilizes similar mapping of V_g and U_g.

Basic principles of hydrodynamic oil and gas entrapment un-doubtedly are sound and correct and techniques for applying them have been well presented. Limitations of the method, however, are not so well understood.

1. A source of continuous water influx must be determined. This is generally presumed to be fresh water infiltration from the outcrops of a continuously permeable formation. This infiltration supposedly occurs toward the center of a sedimentary basin. Nothing is ever said as to what happens to this water once it reaches the center. Perhaps it evaporates and water vapor perco-lates upward; perhaps it continues across the basin and reappears on the other side as springs.

2. If fresh water infiltrates from the outcrops, dilution of for-mation water should be observed and significant dilution patterns around oil and gas fields should be found.

3. The author of the hydrodynamic entrapment theory does not mention the possibility of salt water's flowing from the basin center toward the edges and outcrops, which is a more likely possibility. Hydrodynamics of compaction should therefore be considered.

Procedure for Hydrodynamic Entrapment Mapping

A brief discussion of procedures for hydrodynamic entrap-ment mapping appears through the schematic sequence of maps in Figure 10-1. First, a structure map (Map I) must be made on top of the porosity zone.

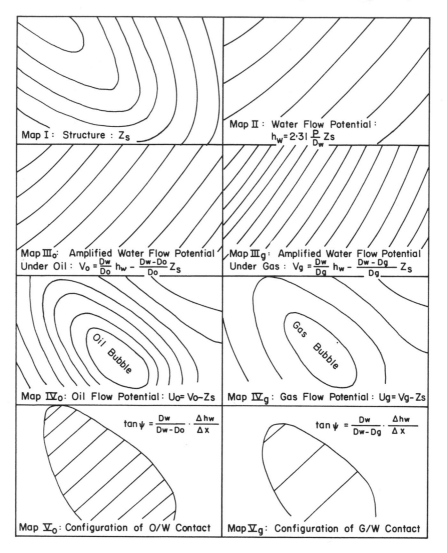

Figure 10-1. Schematic sequence of maps required to delineate hydrodynamic oil or gas entrapment.

Deriving Water Flow Potential: V_o or V_g

Deriving the water flow potential at each well begins with the following information:

1. *Static water pressure* is mostly from pressure measurements during Drill Stem tests as the initial shut-in pressure, but also from pressure build-up tests after production tests and from wire line Formation Tester tests. Where more than one static pressure value results from various tests, the most likely pressure value (often the maximum pressure reached) should be selected.

2. *Depth of the recording pressure gauge* must be accurately recorded.

3. *Formation water density* at the prevailing salinity, temperature and pressure must be specified.

4. *Water flow potential function* (h_w) may then be calculated by

$$h_w = 2.31 \ \frac{P_s}{D_w} \ + Z_g \qquad (10\text{-}7)$$

where:

P_s = static pressure measured, psi

D_w = water density in grm/cm^3.

Z_g is the elevation of the measuring pressure gauge in feet above a reference datum; h_w is the water flow potential or hydrostatic head of reservoir water in feet above a datum level. It is also the height in feet to which a static reservoir water would rise in a casing cemented in the formation and in the absence of expanding gases.

Plotting each well's h_w values and their contouring yield a true water flow potential map of the reservoir water (Map II). However, to facilitate interpretation, it is desirable to amplify the h_w values by factor $D_w/(D_w - D_o)$—where D_o is the oil density under prevailing reservoir conditions of pressure, temperature and gas composition.

Map III_o plots the values of amplified water flow potential (V_o) so obtained at each well. The contour intervals are proportional to the water flow potential gradients, thus indicating the direction of water motion. Contours of the V_o map are actually equipotential lines, and the water flow direction is perpendicular to these lines everywhere and directed from high to low values. A similar Map III_g may be made for the amplified water flow potential $V_g = h_w (\frac{D_w}{D_w - D_g})$.

Estimating water flow velocity in the reservoir which causes the oil-water interface tilt utilizes Darcy's law of fluid flow in porous media for horizontal flow:

$$v = -\frac{K(D)}{\mu_{w\ (cp)}}\ \frac{\Delta h_w\ \text{(atmospheres)}}{\Delta\ x\ \text{(cm)}} \qquad (10\text{-}8)$$

The velocity v (cm/sec) in free space is so obtained.

Deriving Oil Flow Potential: U_o

Subtracting the corresponding structural elevation (Z_s) in feet above the reference datum from the V_o value anywhere on the V_o map gives the oil flow potential (U_o) immediately below the structure's roof. The Z_s values are from a structure (Map I) drawn on the top of the reservoir from seismic and subsurface data of well logs. Maps I and III are then traced on transparent films so that intersections of contour lines can be seen and read easily. The V_o-Z_s difference gives the U_o value, which is eventually plotted on Map IV_o and contoured, thereby giving the equipotential lines of the oil flow potential at the top of the reservoir.

Oil entrapment regions on Map IV_o are areas of minimum oil flow potential delineated by closed contours. The map explains why certain small structures are barren of oil, possibly from being flushed by water. No closed contour for the oil flow potential is at these wells; therefore, no oil accumulation could remain there under hydrodynamic conditions that Map II depicted, even though closed structures are at each well. A similar procedure derives Map IV_g, which is the gas flow potential map.

Determining the Configuration of the Oil-Water Contact

In existing oil fields, the oil-water contact (defined above) can generally be observed directly or may be determined by downward projection from the appropriate well log combination.

Direct observation of the oil-water contact from well logs is by recognizing the minimum depth at which a R_o value is observed. R_o, resistivity at 100 percent water saturation, is at times obvious on logs measuring true resistivity (induction log or laterolog 7 corrected for Delaware effect) when not distorted by invasion. When logged and drilled in salt muds, determining oil-water contact is greatly facilitated by running a laterolog-microlaterolog. Then, observing the minimum depth at which the two curves read the same value (i.e., R_o value) can minimize the effect of invasion.

An independent verification that the tilt is reasonable under hydrodynamic conditions is obtained by the theoretical oil-water contact tilt formula:

$$\tan \phi = \frac{D_w}{D_w - D_o} \cdot \frac{\Delta h_w}{\Delta x} \qquad (10\text{-}9)$$

where:
  ϕ = tilt angle;
  D_w and D_o = respectively, water and oil densities in place
  Δh_w = differential water flow potential between two wells
  Δx = horizontal distance between the same two wells.

Oil-water contacts so determined definitely do not consider the effect of faults that may have been recognized within reservoir structures. To study disturbances that faults introduce into the rather simple oil-water configuration would require working with large and detailed maps in examining fluid equilibrium levels across the sealing and nonsealing faults. Such a study is possible when complete capillary pressure information exists at all wells. By combining the observed oil-water contacts and tilt angle computed from (10-9), it is possible to map the oil-water contact configuration, such as Map VI indicates.

Projective Technique to Find Hydrocarbon-Water Contacts

Water motion in the earth is generally derived from hydrostatic pressure measurements in aquifers associated with prospective oil reservoirs and by converting them to flow potentials. Pressure measurements should be accurate and dependable to a fraction of a pound per square inch.

Intensively exploring for hydrodynamic traps encompasses many inadequacies. The progress in instrumentation improvement here has not kept pace with the development of understanding hydrogeological oil entrapment mechanism. Data necessary for successfully using hydrodynamic principles in the search for oil are therefore often inadequate when available; most of the time, they are unavailable.

To help inadequacies of current hydrodynamic exploration techniques, Pirson proposes to use well logs and, more specifically, modern well log combinations to determine the depth to specified strata under preassigned states of fluid saturation. Most generally, this will be levels of 100 percent water saturation in the different strata of varying petrophysical character.

It must naturally be assumed that fluids in reservoir rocks are in a state of capillary equilibrium or that no short or long-term transients are occurring. Then, water saturation may increase with depth uniformly if the reservoir rock is of uniform lithology. The saturation increase is logarithmic with depth. Hence, it would not be difficult to project this increase downward until a 100 percent water level is attained if it weren't for rocks' never being homogeneous.

Petrophysical consideration and an adequate, secured suite of logs can normalize the calculated water saturations to those that would appear in a reservoir rock of homogeneous and uniform petrophysical properties (porosity, permeability, tortuosity and dispersivity—the four characteristics of an intergranular porous rock). By selecting different normalization bases to correspond to the properties of a reservoir's main zones, the downward projection of the water saturation in each zone helps establish oil-water contact levels per zone.

Such oil-water contacts need not be visible in the logs themselves. A well which is only partially penetrating a reservoir formation can establish the depth of the expected oil-water contact. In other words, to what depth the well should be drilled further can be predicted in exploring fully a reservoir's producing potentialities.

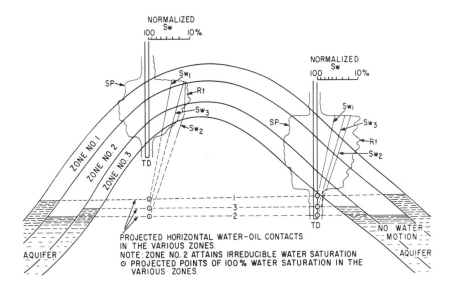

Figure 10-2. Schematic representation of oil entrapment under hydrostatic conditions determinable from well log analysis in a three-zone reservoir in two wells. (Courtesy of *World Oil*.)

If the particular well is drilled at or near the apex of a closed structure or trap, downward water saturation projections may be carried out to 100 percent saturation levels to establish the height of structural closure to be expected in each reservoir zone (Figure 10-2).[9] This information plus prediscovery (geological or seismic) structural maps help establish the accumulation's areal extent. This is extremely important in acquiring additional promising acreage, dealing with offset lease owners, planning the next well and developing the field in an orderly fashion.

As new wells are drilled, applying projective water table level determination techniques helps verify if the oil-water contact is flat, indicating thereby the absence of water motion under the hydrocarbon accumulation; or whether it is tilted, thus indicating an entrapment under hydrodynamic conditions. Mapping the intersection of the tilted water table with the structural trap, as geological and/or geophysical considerations have determined, will

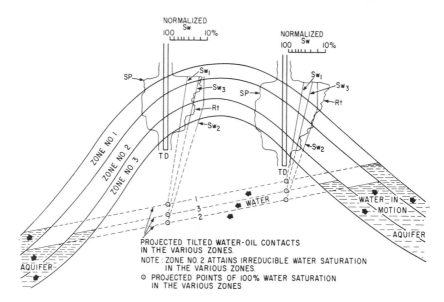

Figure 10-3. Schematic representation of oil entrapment under hydrodynamic conditions determinable from well log analysis in a three-zone reservoir in two wells. (Courtesy of *World Oil*.)

permit the oil operator to evaluate how much of the structure on the down-dip side of the tilted oil-water contact may be expected to be productive. This again gives this informed operator a decided advantage in dealing with neighboring landowners and operators (Figure 10-3).[9]

In fields already developed, acquiring promising but unleased outlying acreage down-dip from the structure's crest may be made with considerably greater certainty of discovering substantial peripheral production. For this purpose, properly combining logs from three wells that penetrate the reservoir rock defines the position and tilt of the oil-water contact in space, even though neither of these wells reach the oil-water contact proper.

Additional important information may be derived with respect to levels of well completion, since the state of fluid saturation at any level in specific zones of the reservoir rock can be established. The reservoir's zonation may be based on well logs by SP and resistivity curve deflections. To avoid water production from such zones, they should be perforated at levels where irreducible water saturation is achieved.

Within a seemingly continuous but stratified reservoir, various zones may not be in capillary equilibrium. This is probably because a shale barrier that prevents capillary equilibrium separates the zones, although the individual zones may be connected to and are in capillary equilibrium with their respective independent aquifers. The lack of capillary equilibrium between zones of an apparently continuous reservoir may also result from the reservoir's being cut by a sealing fault across which the fluids are prevented from coming into capillary equilibrium.

When a reservoir rock's lithology highly varies from rapid lateral variations in the sedimentation, techniques discussed herein help establish whether the sands suspected of being contemporaneously deposited have the same aquifer when—otherwise—the value of their correlation by log character, by recognizing sequences of reproducible events or by fossils is doubtful. Establishing that all sands have a common aquifer may thus prove correlation and reservoir identification.

Example A

Figure 10-4[9] logs are from the Wilcox sands in South Texas and comprise an induction-electric and a sonic log. Two sands appear for which the usual techniques calculated and normalized water saturations to a 27 percent porosity zone. When plotted on a logarithmic scale, water saturations are projected to oil-water contacts at 7,622 and 7,781 feet, respectively. This indicates that the sands are associated with independent aquifers and that the reservoir closure is limited to the effective pay of the sands unless structurally higher positions may still be drilled on the structure or trap.

Example B

Figure 10-5[9] logs are from the Deep Edwards limestone in South Texas; they comprise an induction-electric and a sonic log. Various zones are distinguishable within this thick reservoir. Water

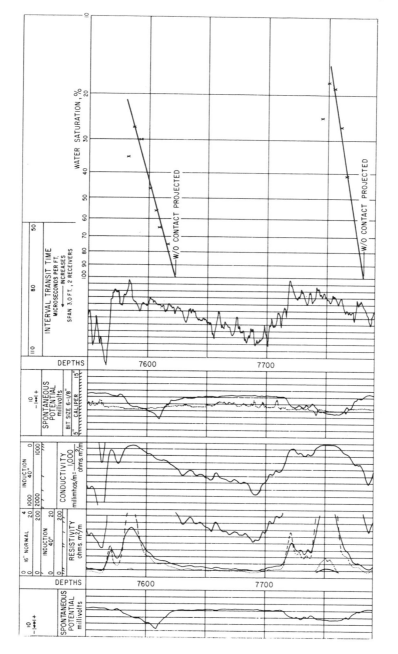

Figure 10-4. Water-oil contact levels obtained in Wilcox sands (South Texas well) by downward projection of water saturation on a log scale. (Courtesy of *World Oil*.)

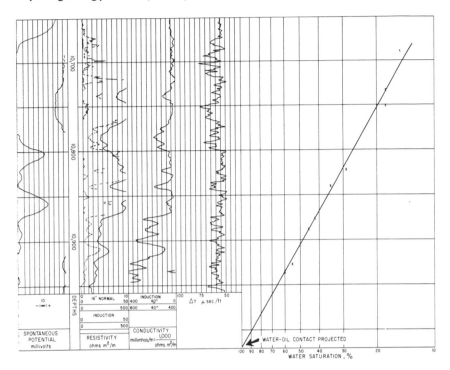

Figure 10-5. Water-oil contact level obtained in the Edwards lime (South Texas well) by downward projection of water saturation on a log scale. (Courtesy of *World Oil*.)

saturations calculated at levels shown are normalized to a 17 percent porosity and fall on a reasonably straight line when plotted on a logarithmic scale. This straight line projects downward to an oil-water contact at 11,018 feet. All porous zones of this reservoir are therefore associated with a common aquifer over which each zone's reservoir closure is readily evaluated. This reservoir's gross closure is 350 feet, assuming that structurally higher positions no longer exist on this structure.

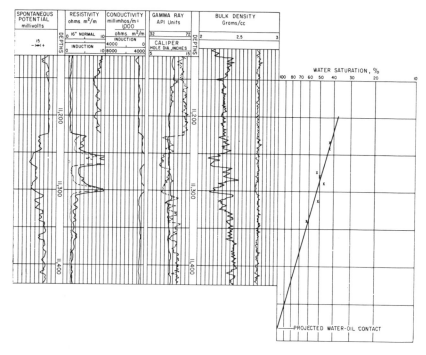

Figure 10-6. Water-oil contact level obtained in a South Louisiana well by downward projection of water saturation on a log scale. (Courtesy of *World Oil*.)

Example C

Figure 10-6[9] logs are from South Louisiana and include an induction-electric and a formation density log. Various zones are distinguishable within this reservoir. Water saturations calculated at different levels and normalized to a constant 21 percent porosity fall on a reasonably straight line when plotted on a logarithmic scale. This line projects downward to an oil-water contact at 11,383 feet. This reservoir's porous zones are associated with a common aquifer over which the total reservoir closure at this well is 150 feet.

Required Log Combinations

The main requirements for a log combination's suiting the interpretation discussed thus far are that the logs should have a sufficiently fine vertical definition and at least one porosity tool of similar or better vertical resolution. Such combinations follow:

Induction-electric log plus sonic log;

Induction-electric log plus density log;

Induction-electric log plus neutron log;

Induction-electric log plus any combination of the above porosity logs;

Laterolog 3 and 7 plus any combination of the above porosity logs;

"Little slam" combination: laterolog 8, induction log 5 FF 40 and 6 FF 40 and a porosity log (sonic or density);

"Grand slam" combination: proximity log, laterolog 8, two induction logs and a porosity log.

The conventional logs (two normal and one lateral curve) do not suit for this work due to their geometric distortion and lack of vertical definition, unless a complete core analysis also exists. The field interpreted log, such as the FAL (which records the R_{wa} curve), highly suits this work after proper modifications and reinterpretation.

Example of Application

Mapping Problem 5
Projection to Oil-Water Contact

A log from a deep well on the Gulf Coast is in Figure 10-7a. It shows individual massive sands separated by shale breaks and it is desired to establish (1) oil-water contact levels; and (2) whether the sands have a common aquifer and what their levels are.

To carry out these calculations, it is necessary to "zone" the various sands, i.e., to separate them into individual members that might have substantially the same physical characteristics within a

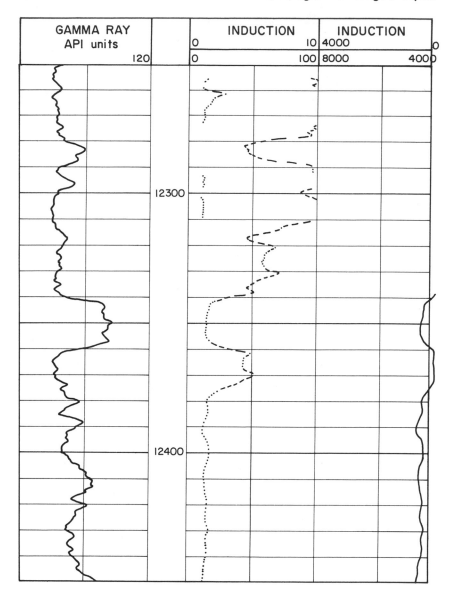

Figure 10-7,a. Example of log combination (gamma ray, induction, sonic) suitable for projection interpretation to the water-oil contact.

narrow range. The controlling characteristics might be porosity and absence of shale. SP and caliper curves may thus be used to advantage with the sonic log (or any other porosity tool).

Several methods may establish the oil-water contact in each zone, but one or two may be more appropriate under certain circumstances. For instance, these conditions may exist offshore and, particularly, in the North Sea (where wells are drilled with salt saturated muds and where the laterolog-microlaterolog has to be used when a transition zone is well defined on the laterolog, the uppermost level at which $R_{LL} = R_{MLL}$ is taken as the oil-water contact). It is, however, not always so easy to find the uppermost level where these two resistivities are substantially equal.

The oil-water contact may also be found by projecting the resistivity gradient downward on the laterolog until the computed or observed R_o value is obtained. Considering the section's lithology and porosity variations, this method is rather questionable. It is necessary to allow for lithologic variations with depth, which may be done according to the following reasoning.

In the homogeneous reservoir rock, Tixier has shown that the true resistivity (R_t) in the capillary transition zone above the oil-water contact, or highest level of 100 percent water saturation (R_o), is a straight line function of depth:

$$R_t = R_o + a (h_o - h) \qquad (10\text{-}10)$$

where:
  h = variable depth
  h_o = depth at R_o (the largest number)
  a = slope of the straight line.
Since $S_w = \sqrt[2]{R_o/R_t}$, squaring both sides of Archie's formula,

$$S_w^{-2} = \frac{R_t}{R_o} = 1 + \frac{a}{R_o} (h_o - h) \qquad (10\text{-}11)$$

which shows that plotting S_w^{-2} versus depth should give a straight line. However, this involves many tedious calculations.

Observing from the R_{wa} method that $S_w = \sqrt{R_w/R_{wa}}$ and substituting gives

$$\frac{R_{wa}}{R_w} = 1 + \frac{a}{R_o}(h_o - h) \qquad\qquad (10\text{-}12)$$

which ratio is also a straight line versus depth.

The R_{wa} function is already in many computed modern logs. Hence, one may plot R_{wa} directly versus depth on the log and observe the points on straight line segments. They are then projected to the R_w value, and individual oil-water contacts are thereby defined. This technique allows to a certain extent for lithology variations with depth. A more satisfactory technique would require normalizing water saturations to a constant porosity and permeability rock.

As an example, the logs of Figures 10-7a and 7b, comprising an induction electric and a sonic log, are used. Readings of R_t and Δt are every 2 feet deep (Table 10-1), and the porosity and formation factor (F) are also calculated and entered in the table. The values of $R_{wa} = R_t/F$ are calculated in the last column. Figure 10-7c plots them, showing four oil-water contacts. This indicates that the sands are independent of each other and do not have a common aquifer.

Hydrodynamic Flushing

In regions where fresh water infiltrates at the outcrops of continuously porous and permeable rocks (as in the Denver-Julesburg, Powder River, Delaware and San Juan sedimentary basins), oil fields—i.e., those where the oil has not been flushed—have water resistivities of lesser magnitude than within the same reservoir outside the confines of the oil accumulations. This is generally attributed to flushing; but, as is seen later, the salinity difference could be from electro-osmosis and the building-up of an osmotic pressure barrier between the fresher water around and the saltier water within the reservoir rock.

(Text continued on page 309.)

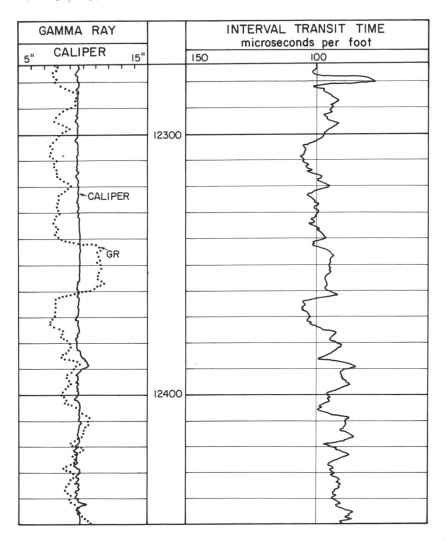

Figure 10-7,b. Example of log combination (gamma ray, induction, sonic) suitable for separate water-oil contacts in independent reservoir zones.

Table 10-1

Answer to Mapping Problem 5 — R_{wa} Method (Example Log 10-7a and 7b)

$R_w = 0.025 @ 209°F$

Depth	$R_t:\Omega m$	$\phi:\%$	F	$R_{wa} = \dfrac{R_t}{F}$
12,270	12	23.6	13.5	0.89
72	11	28.4	9	1.22
74	10	28.3	9	1.11
76	10	20.5	20.5	0.52
78	9	11.9	61	0.15
80	6	23.5	13.5	0.44
82	4	16.4	30	0.13
84	5	16.1	31	0.16
86	5	17.4	26	0.19
88	7	13.9	35	1.11
90	10	23.3	14	0.71
92	10	24.0	13	0.77
94	12	23.3	14	0.86
96	11	12.4	54	0.20
98	11	25.3	12	0.92
12,300	9	27.8	9.6	0.94
02	10	26.7	10.5	0.95
04	11	29.4	8.8	1.25
06	12	29.3	8.8	1.36
08	11	26.6	10.5	1.05
10	11	28.1	9.4	1.17
12	8	27.5	9.7	0.82
14	7	23.9	13	0.54
16	5	24.4	12.5	0.40

Table 10-1 (continued)

18	5	15.8	33	0.15
20	6	26.9	10.5	0.57
22	6	27.6	9.5	0.63
24	6	23.6	14	0.43
26	6	24.3	12.5	0.48
28	6	25.1	12	0.50
30	7	26.9	10.5	0.67
32	6	27.5	9.5	0.63
34	5	27.2	9.4	0.53
36	5	23.9	13	0.38
38	5	23.9	13	0.38
40	3	23.5	13.5	0.22
42	2	10.3	80	0.02
44	2	9.1	102	0.01
46	2	9.7	93	0.01
48	1	9.7	93	0.01
50	1	8.4	125	0.01
52	1	10.3	80	0.01
354	1	8.3	125	0.01
56	2	10.2	80	0.02
58	2	10.3	80	0.02
60	3	13.8	44	0.07
62	5	29.4	8.2	0.61
64	4	29.3	8.2	0.49
66	4	28.0	9.4	0.43
68	4	22.7	14.5	0.28
70	5	28.8	9	0.56
72	4	26.5	10.5	0.38
74	3	20.2	19	0.16
76	2	19.3	21	0.10

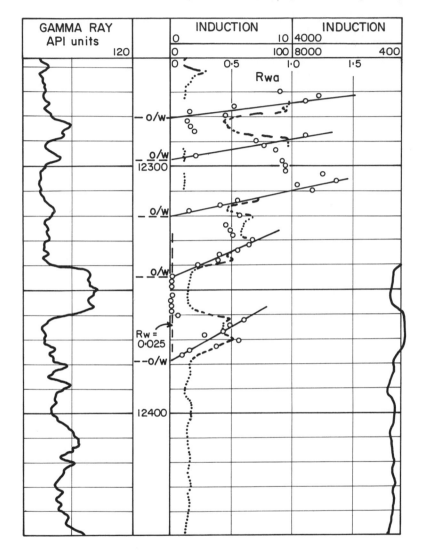

Figure 10-7,c. Solution of the log combination in Figures 10-7,a and b, indicating five separate water-oil contacts in independent reservoir zones.

Observing true hydrodynamic flushing around an oil accumulation rests on these requirements:

1. The reservoir rock must be a continuous permeable bed with few inhomogeneities; this requirement is seldom satisfied. The most likely formations in which this requirement may be satisfied are the quartzose rocks.

2. A permeability barrier must protect the oil field from flushing; this could be a pressure barrier. Such permeability barriers could result from faulting, or from a permeability seal, like clay swelling under fresh water invasion may produce. This is believed to occur in the San Juan Basin gas fields.

3. Influx of water must require that the hydrostatic pressure of the infiltrating water be greater than formation pressure. Again, an osmotic pressure barrier more readily explains this requirement.

4. Formation dip from outcrops toward the basin center must be sufficient to allow fresh water infiltration; it need not be more than one or two degrees. This calls for fluid moving opposite the direction which is expected at the dominant direction resulting from basin compaction.

While true hydrodynamic flushing by fresh water appears unlikely after oil entrapment has occurred, computing the formation salinity from the SP curve may verify its existence.

Examples of hydrodynamic flushing appear in the literature. One of the earliest studies was over the Salt Creek Field, Wyoming, in the "First Wall-Creek sand," a faulted-anticlinal structure west of the Powder River Basin (Figure 10-8). The sand outcrops 30 to 50 miles west in the south extension of the Big Horn Mountains, from which it is natural to expect surface waters of low salinity actively infiltrating east toward the center of the Powder River Basin. Contours lines of equal chloride content show a high value of 50.0 for the chloride reacting value (i.e., $\frac{1}{MW \times n}$ x ppm Cl, approximately) toward the structure's apex with values as low as 5.0 on the west, or infiltrating side, and 9.0 on the east, where flushing has not been as effective.

Jones[8] has published other examples of seemingly true hydrodynamic flushing over these fields in Wyoming: Miller Creek,

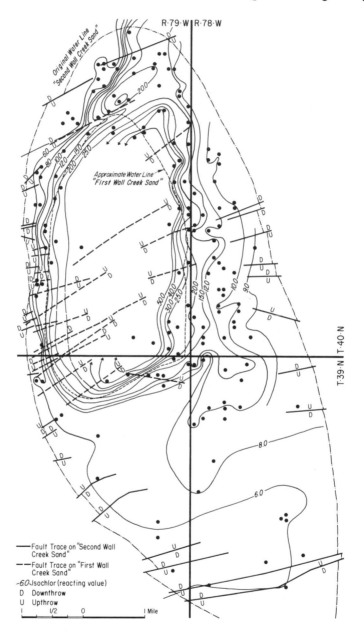

Figure 10-8. Isochloride contour lines of the formation water in the "First Wall Creek Sand" of the Salt Creek Field, Wyoming. (Courtesy of W.T. Thom, Jr., E.M. Spieker and the U.S. Geological Survey.)

Coyote Creek, Kummerfeld, Donkey Creek and Halverson. These are also in the Powder River Basin, where infiltration water seems to be most abundant. However, in these cases, the oil fields are to the east of the basin, where the flushing comes from.

Flushing direction is determined by salinity shadow effects produced by oil accumulations that prevent fresh water dilution from affecting formations down-dip. Figure 10-9[8] well illustrates this between the Miller Creek and the Kummerfeld fields. The dilution effect seems to fan-out after it has passed between the

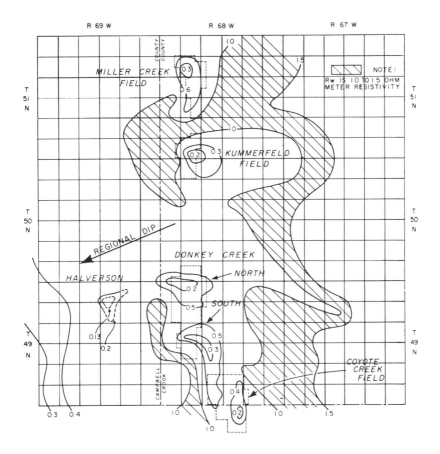

Figure 10-9. Dilution patterns of formation water in the Dakota sand (Crook and Campbell counties, Wyoming) as derived from water resistivities. (Courtesy of R. E. Jones and *The Oil and Gas Journal*.)

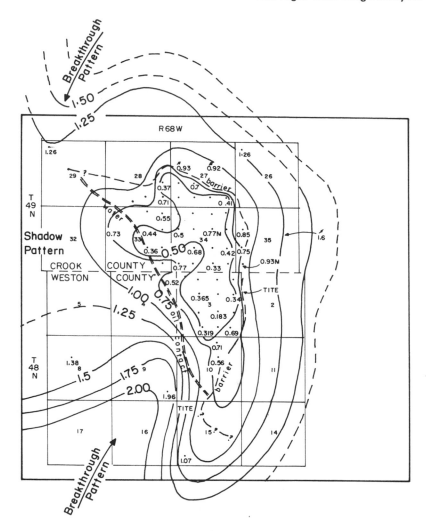

Figure 10-10. Dilution patterns of Dakota formation waters in and around Coyotte Creek Field, Wyoming. (Courtesy of R. E. Jones and SPWLA.)

two fields. Accordingly, dilution patterns may be greatly significant in suspecting an oil accumulation, and the contouring of isosalinity lines should be carefully examined for possible breakthrough patterns between fields and salinity shadow patterns down-dip from the field. Such patterns (Figures 10-10, 10-11) seem to be in the Denver-Julesburg Basin between the Adena and

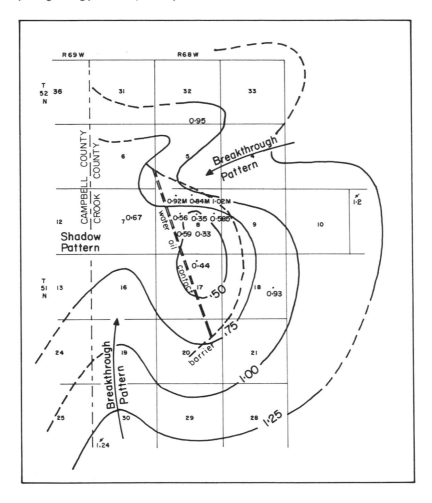

Figure 10-11. Dilution patterns of Dakota formation waters in and around Miller Creek Field, Wyoming. (Courtesy of R. E. Jones and SPWLA.)

Leader fields and apparently can help detect oil accumulations there.

As Jones has pointed out, favorable water salinity patterns are insufficient to warrant the existence of an oil field, and the water saturation should also be mapped in those sands. If water saturation shows a decreasing trend toward the center of the shadow pattern, this should be further evidence of an oil accumulation.

Determining Fresh Water Flushing From Logs

The method is simple and mainly determines R_{we} from the SP curve. Therefore, the porous rock involved must be relatively clean and devoid of clays reactive to fresh mud exposure. If the sand is shaly, one should probably look for a shalyness or permeability barrier with the SP plot or for an osmotic pressure barrier. The SP curve should be corrected for the usual factors—bed thickness, true resistivity, borehole effects, invasion, etcetera.

It's more important, however, to check whenever possible against other sources of water salinity and resistivity information such as (1) water samples from drill stem tests, producing wells, etc. with the appropriate safeguards to ensure representative samples; (2) calculating R_w from the sonic log and density log in shales adjacent to the particular sand.

Using shale water resistivity assumes that adjacent sands and shales have the same water salinity in the absence of hydro-dynamic flushing. If a sonic log in shale is available, from Δt_{sh} a shale porosity (ϕ_{sh}) and a shale formation factor may be calculated.

$$F_{sh} = 0.8\ \phi^{-2} \text{ or } 0.62\ \phi^{-2.5} \qquad (10\text{-}10)$$

Thus:
$$R_w = R_{sh}/F_{sh} \qquad (10\text{-}11)$$

where R_{sh} is obtained from a short normal curve.

However, this is the same value as R_{wa} in shale which has been consistently three to five times true R_w in sands, unless fresh water infiltration has flushed the sand. This is thus another way of checking on flushing. If a density log is available, then

$$\phi_{sh} = \frac{2.65 - D_{sh}}{2.65} \qquad (10\text{-}12)$$

where D_{sh} is the measured shale density. F_{sh} and R_w are then calculated as above by (10-10) and (10-11).

These calculations of R_w from shale characteristics preassume that no osmotic fluid has been exchanged between sand and shale, i.e., that they are substantially at the same pore pressure. Hydrodynamic oil entrapment and flushing result from the same phenomenon—namely, fresh water infiltrating from the outcrops toward the basin center; this requires unusual uncommon conditions found particularly in the Rocky Mountain basins.

Example of Application

Mapping Problem 6
Hydrodynamic Flushing

This problem used well logs in Mapping Problem 2 from the Black Jack Field in the Denver-Julesburg Basin. Here oil production is generally in the D and J sands; but, because they outcrop to the east, as well as in the Rocky Mountain foothills, some geologists believe that fresh water infiltration might have flushed oil from one or both sides. If so, the connate water should have become diluted, and characteristic breakthrough and shadow dilution patterns should have developed around probable oil accumulations. Mapping Problem 6 seeks such possibilities.

Examples of the required computations are in Table 10-2 for some logs from the area (shown in Figure 1-3). SP readings have

Table 10-2
Computation for Mapping Hydrodynamic Flushing and Hydro-osmotic Entrapment

Well Number	Interval	SP mv	Rm Ωm	°F	R_{mf} @Tf	R_{mf}/R_{we}	R_{we} Ωm	Osmotic Pressure psi
1	5375-400	-30	1.4	126	1.15	2.5	0.46	1800
2	5394-430	-20	1.6	136	1.35	1.8	0.75	very high
3	5285-318	-20	1.3	131	1.00	1.8	0.55	2900
4	5285-312	-20	1.1	137	0.85	1.8	0.47	1900
8	5410-455	-40	2.0	125	1.70	1.7	0.52	2500

been made in the second bench of the J sand. The R_{mf} values were computed from mud resistivity values off the log readings. R_{mf}/R_{we} values were computed from the static SP equation or from an SP salinity chart (Schlumberger Chart A-10).

R_{we} values so obtained were plotted at the Figure 10-12 wells, which show certain flushing around the Black Jack Field and another flushing effect around the North one-half of Section 29.

Hydro-osmotic Studies in Thin Shaly Sands

Using well logs as hydrodynamic tools to find oil-water contact is satisfactory when studying relatively thick reservoirs with modern well logs, which have a fine vertical resolution (induction logs or laterologs), plus a porosity log of equal or better resolution (sonic, density or neutron in clean rocks). In thin reservoir rocks, the downward projection to an oil-water contact is no longer possible. This is also impossible with conventional logs, especially when porosity tools of only poor resolution are available.

When sands are shaly, however, it is possible to take advantage of the osmotic flow of formation water through partially semi-permeable membranes (the shaly sands) to establish whether the formation water underlying an oil field is moving and its direction. This is especially significant in searching for stratigraphic traps which are generally associated with shaly sands because a commercial accumulation then is usually not discovered unless a down-dip water motion can be established. This is because in shaly sands the feather-edge or pinch-out of a stratigraphic trap does not provide an adequate capillary barrier to retain a commercial oil column under hydrostatic equilibrium. It would be worse yet if the formation water flow should be up-dip toward the pinch-out.

Hydrodynamic flow in geologic sections of a stratigraphic trap character is unlikely, since it requires the Darcy type flow. However, chemi-osmotic flow, as well as thermo and electro-osmotic flow, are entirely possible. It is extremely important to determine if they occur in the proper direction to establish an adequate commercial oil entrapment. The basic principle of this approach to hydrodynamics is that shales and shaly sands, being semipermeable

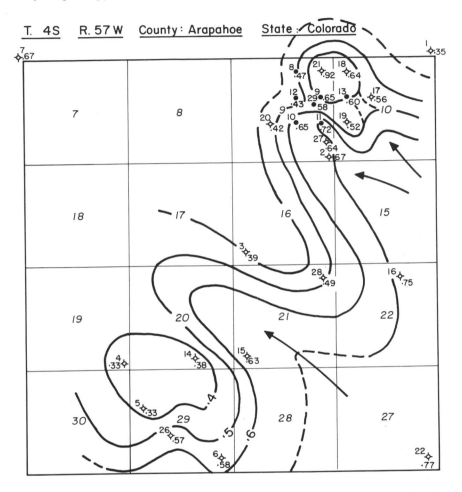

Figure 10-12. Example of dilution pattern mapping from electric logs in and around the Black Jack Field, Colorado.

membranes to ions in solutions of different ionic concentrations, will be under different pore pressures.

Osmosis is the spontaneous flow of solvent only through a semipermeable membrane, i.e., permeable only to the solvent and not to the solute or dissolved solids. A semipermeable membrane is not a permoselective or an ion exchange membrane. Osmosis causes fishes at sea to have body fluids at a lesser salinity than sea water, sap in trees to rise to greater heights than sizes of the capillaries predict, dry seeds to swell in water, etcetera. The theoretical pressures osmosis exerts are tremendous; van't Hoff's equation for dilute solutions predicts them:

$$\pi = RTC \tag{10-13}$$

where:

   π = osmotic pressure, atmosphere
   R   = 8.31, joule/°K
   T   = temperature, °K
   C   = solute concentration, mole/liter.

For a concentration difference of one mole of NaCl per liter in water (58,000 ppm) at 100°C, an osmotic pressure of nearly 3,100 atmospheres is computed. This pressure is the excess pressure of the dilute solution over the concentrated solution, which results in a spontaneous diffusion of fresh water into the more concentrated solution. Ostwald has demonstrated that osmotic pressure is independent of the nature of the membrane. All semipermeable membranes give the same osmotic pressure provided the membrane does not leak or break. The nature of the membrane and its area determine how rapidly osmosis occurs. The osmotic pressure computed is also independent of the solution and the solvent.

Shales and shaly sands may be considered as semipermeable membranes, however imperfect. Compacted clays behave as ideal semipermeable membranes because, when filtering NaCl solutions through them, they exclude chloride ions completely from the internal solution regardless of the salinity of the solution in which the membrane is placed. Shales in the earth act as ion filters and lead to hydrodynamic desalting of moving ground waters.

The mechanism by which osmotic pressure differences are created in the earth between shaly-silty formations and cleaner formations may occur in three stages.

Stage 1 is suspension period of clay and sand particles in sea water before sedimentation occurs. Clay particles act as micelles (M⁻), carrying net negative charges because of the preferential absorption of Cl⁻. In the immediate vicinity of the clay micelles, an "internal solution," wherein the micelles [M⁻] and ions [Na⁺]

are in molal equilibrium, may appear. This solution is in contact with an "external" solution containing only Na^+ and Cl^- in solution. For equilibrium between the two solutions, the following relation must exist:

$$(M^-) \cdot (Na^+) = (Cl^-) \cdot (Na^+) \qquad (10\text{-}14)$$

Stage 2 evolves when sedimentation of the previously suspended clay particles has occurred and the clay micelles have therefore lost their mobility. However, because of their negative charges, they attract and hold in their neighborhood—within the internal solution—a certain concentration (x) in Na^+, which an equal concentration (x) in Cl^- must neutralize. These ions of concentration x come from the external solution; and, if the original concentrations of micelles and ions were considered as molal, the following equation may be written, expressing the Donnan equilibrium:

$$x(1 + x) = (1 - x) \cdot (1 - x) \qquad (10\text{-}15)$$

where:
  x $= Na^+$ in the internal solution
  $1 + x = Cl^-$ in the internal solution
  $1 - x = Na^+$ and Cl^- in the external solution.
solving equation (10-15) gives x $= 1/3$.

Hence, the external solution—the one communicating with the clean part of a sand which grades laterally into a shaly sand—is depleted of one-third its original salt content. However, on the shaly side, or internal solution, most ions are immovable; and the free internal solution is only half as salty as the external solution. This important conclusion explains why water samples on DST from shaly sections are less salty than corresponding samples from clean sands. The above computed salinity contrast of 1/3 is for a perfect membrane. Shaly sands would show a lesser contrast.

Stage 3 explains that, with Donnan equilibrium, ionic transfer from the external to the internal solution can no longer occur; and the separation between the two solutions may be considered as a semipermeable membrane through which only water may flow.

The third stage is thus that in which water diffusion from the low concentration (shaly sand) toward the high concentration (clean sand) tends to occur; i.e., osmosis, which forms a pressure barrier between clean sand and shales and shaly sands, occurs. The pressures are, therefore, greater in the pore space of shales and of shaly sands than in their stratigraphically equivalent clean sands in a closed system not leaking and where all fluids are in equilibrium. If gas, oil and water should be entrapped in this closed (structural or stratigraphic) system, an apparent tilt of the hydrocarbon-water contact will be away from the shaly section, even though fluid no longer moves under osmostatic equilibrium.

Shales and shaly sands may not be considered as perfect membranes. Accordingly, theory may not predict differences in osmotic pressure. To predict the pressure of an osmotic pressure barrier from shalyness determination, observed pressures from Formation Tester, Drill Stem (DST) tests or production tests versus measured formation shalyness factors must be calibrated. Well logs and well log combinations that may determine these may include (1) *SP curve* through the α shalyness factor or the K constant or the R_{we} calculations; (2) *gamma ray curve* through the γ shalyness factor; (3) *density-sonic* log combination through the q shalyness factor, which measures the pore space filling in colloidal or dispersed clay; (4) *density-neutron* log combination through the p shalyness factor, which measures the gross pay section that shale laminations occupy.

An example of calibration of the osmotic pressure difference by known differential pressure measurements, such as those made during DST at the initial shut-in pressure (SIP), was over a known field in the Denver-Julesburg Basin, Colorado. Figure 10-13[9] represents the well locations and results of DST initial shut-in pressure measurements. Figure 10-14 cross plots SIP versus R_{we} off the SP curve. Figure 10-15[9] depicts the calculated differential pressures from the indicated brine resistivities as computed from the SP curve. Considering that pressure measurements lack much accuracy (i.e., in even picturing closely the actual bottom hole pressures), correspondence between the two pressure maps is striking. Figure 10-15 indicates that the oil entrapment is by an

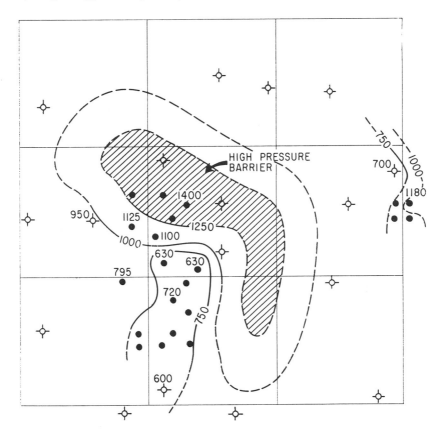

Figure 10-13. Observed shut-in pressures in psi on drill Stem Tests of J sand in Denver-Julesburg Basin area. (Courtesy of *World Oil*.)

osmotic pressure barrier to the east and up-dip from the oil accumulation in this J sand field.

From pressure testing in wells, it would thus appear that reservoir fluids (namely, water) move down structure, possibly providing a taller oil column that a capillary pressure barrier to the east could not entrap. An osmotic pressure barrier is supposedly a more realistic oil entrapment phenomenon than expecting hydrodynamic fluid motion in the Denver-Julesburg Basin (as Russell[12] proposes).

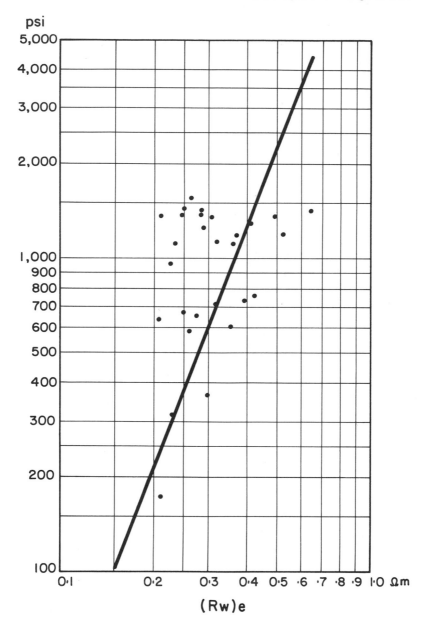

Figure 10-14. Calibration of osmotic pressure measured by DST in the J sand, Denver-Julesburg Basin.

In the example field under discussion, the entrapment mechanism is completely devoid of structural elements except for the regional west dip on the eastern shelf of the Denver-Julesburg Basin. However, structurally entrapped oil in this basin does not necessarily require a pressure barrier to confine the oil.

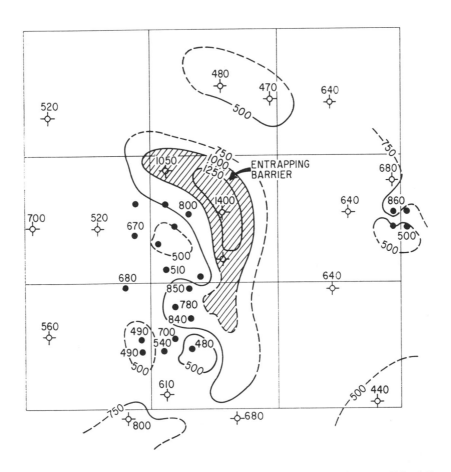

Figure 10-15. Computed osmotic pressure map illustrating possible oil entrapment by an osmotic pressure barrier in the same area as Figure 10-13. (Courtesy of *World Oil*.)

Example of Application

Mapping Problem 7
Hydro-osmotic Oil Entrapment

The well logs and map in Problem 2 will be used again in this problem.

Instead of flushing by fresh water, a water salinity anomaly may develop near an oil accumulation by osmotic fluid transfer, particularly when the potential reservoir rock is shaly as it is here.

To detect an osmotic pressure barrier, the SP deflection versus the recorded formation pressures observed on DST, formation testers or pressure build-up tests must be calibrated. This must be done with great selectivity, as DST measurements are quite inaccurate. The Figure 10-14 calibration chart was obtained this way in the J sand from a nearby field. Thus, it is expected to apply to the region under study.

Map the indicated osmotic pressure and determine whether a pressure barrier that delineates a probable oil field might exist. The procedure consists in determining R_{we}, as in Mapping Problem 6; and Table 10-2 has already recorded the results. From the values of R_{we}, Figure 10-14 gave the osmotic pressure values.

Figure 10-16 plots the osmotic pressure values and shows that the Black Jack Field is a region of low pressure surrounded by pressure barriers except to the southeast. The north half of Section 29 appears again as a favorable area of oil entrapment.

References

1. De Sitter, L. V., "Diagenesis of Oil-field Brines," *AAPG Bulletin*, 31, No. 11, (1947), pp. 2030-2040.

2. Hanshaw, B. B. and E-An Zen, "Osmotic Equilibrium and Overthrust Faulting," *Bulletin Geological Society America,* 76, No. 12, (December, 1965), pp. 1379-1386.

3. Hill, G. A. et al., "Reducing Oil-finding Costs by Use of Hydrodynamic Evaluations," *Economics of Petroleum Exploration,* Englewood Cliffs, N. J.: Prentice Hall, 1961, pp. 38-69.

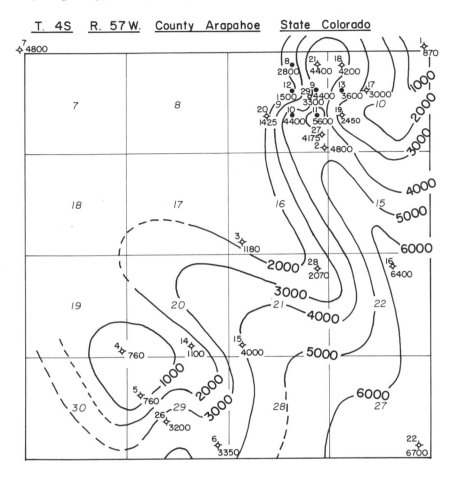

Figure 10-16. Example of osmotic pressure barrier mapping from electric logs in and around the Black Jack Field, Colorado.

4. Hubbert, M. King, "The Theory of Ground Water Motion," *The Journal of Geology*, 48, No. 8, (1940), pp. 785-944.

5. _____, "Entrapment of Petroleum Under Hydrodynamic Conditions," *AAPG Bulletin*, 37, No. 8, (August, 1953), pp. 1954-2026.

6. _____, "Application of Hydrodynamics to Oil Exploration," Paper No. RPCR, 4, Seventh World Petroleum Congress, Mexico City, 1967, p. 33.

7. Jones, R. E., "The Status and Future of Electrical Reservoir Evaluations Using Well Logs," Paper No. 875, 12, A, API Rocky Mountain Meeting, Denver, Colorado, April, 1958.

8. _____, "Electric Log Evaluation in Exploring for Rocky Mountain Petroleum Traps," *The Oil and Gas Journal*, Vol. 58, No. 29, (July 18, 1960), pp. 142-145 and Vol. 58, No. 30, (July 25, 1960), pp. 294-298.

9. Pirson, S. J., "How to Use Well Logs to Seek Hydrodynamically Trapped Oil," *World Oil*, Vol. 164, No. 5, (April, 1967), pp. 100-106.

10. Roach, J. W., "How to Apply Fluid Mechanics to Petroleum Exploration," *World Oil*, Vol. 160, No. 4, (March, 1965), pp. 71-75 and Vol. 160, No. 5, (April, 1965), pp. 131-134.

11. Russell, W. L., "Tilted Fluid Contacts in the Mid-Continent Region," *AAPG Bulletin*, 40, No. 11, (November, 1956), pp. 2644-2688.

12. _____, "Reservoir Water Resistivities and Possible Hydrodynamic Flow in Denver Basin," *AAPG Bulletin*, 45, No. 12, (December, 1961), pp. 1925-1940.

13. Schwab, R., "Logging Important Aspect of Hydrodynamic Studies," *Oilweek*, Vol. 16, No. 25, (August 16, 1965), pp. 36-37.

14. Smith, D. A., "Theoretical Conditions of Sealing and Non-Sealing Faults," *AAPG Bulletin*, 50, No. 2, (February, 1966), pp. 363-374.

15. Summerford, H. E., "Inclined Water Tables," *Wyoming Geological Association Guide Book*, Seventh Annual Field Conference, 1952, pp. 98-102.

Appendix 1

**FORTRAN II Language Program for Calculating
the Dip Magnitude and Azimuth and
the Degree and Orientation of
the Resistivity Anisotropy**
(Courtesy of A. R. Rodriguez)

Nomenclature

Setting up a computer program in FORTRAN language for calculating the dip magnitude and azimuth and the degree and orientation of the resistivity anisotropy, used the following nomenclature. The FORTRAN nomenclature, its equivalent in the theory's nomenclature and a description fall below.

FORTRAN II	Equivalent	Description
AEE	a_E	Component of unit vector $\bar{a}$ along axis OE.
ALPHA	α	Angle between axis OU^1 and the projection onto a horizontal plane of unit vector $\bar{d}$.
ANISOT		Eccentricity of the anisotropy ellipse diminished by one.
ANN	a_N	Component of unit vector $\bar{a}$ along axis ON.
AZIM 1	μ	Azimuth of electrode 1, degrees.

329

AZIMUTH		Dip azimuth relative to true north.
CAPOME	Ω	Orientation of the large semi-axis of the anisotropy ellipse relative to magnetic north.
C1, C2, C3	C_1^2, C_2^2, C_3^2	Reciprocals of R_1^2, R_2^2, R_3^2, respectively.
DECLIN		Magnetic declination, degrees. It is positive when it is toward east.
DEPTH		Depth of the level of interest, feet.
DIA		Borehole diameter, inches.
DIP	Θ	True dip angle.
DRIFT	δ	Drift angle of the hole, degrees.
END	n_D	Component of unit vector $\bar{n}$ along axis OD.
ENE	n_E	Component of unit vector $\bar{n}$ along axis OE.
ENEI	n_A	Component of unit vector $\bar{n}$ along axis OA.
ENEPH	n_F	Component of unit vector $\bar{n}$ along axis OF.
ENN	n_N	Component of unit vector $\bar{n}$ along axis ON.
ENV	n_V	Component of unit vector $\bar{n}$ along axis OV.
ETA	η	Angle between the large semi-axis of the anisotropy ellipse and axis OT.
GAMMA	γ	Angle between axes ON and OU.[1]
H2	h_2	Displacement, parallel to the hole axis, between correlative points on resistivity curves 1 and 11, inches. It is positive when the point on curve 11 is above the point on curve 1.

H3	h_3	Displacement, parallel to the hole axis, between correlative points on resistivity curves 1 and 111, inches. It is positive when the point on curve 111 is above the point on curve 1.
OMEGA	ω	Angle between axis OT and unit vector $\overline{R}_1$.
ORIENT		Orientation of the large axis of the anisotropy ellipse relative to true north.
PHI	Φ	Dip aximuth relative to magnetic north.
PHIB	$\overline{\phi}$	Angle between the large semi-axis of the anisotropy ellipse and the direction of electrode 1 on the bedding plane.
PSI	ψ	Angle between axis OT and the projection onto a horizontal plane of unit vector $\overline{R}_1$.
RVBRG	β	Relative bearing, degrees.
R1, R2, R3	R_1, R_2, R_3	Correlative resistivity values read on resistivity curves 1, 11, and 111, respectively.
SIGMA	σ	Angle between axes OT and ON.
TGALPH	$\tan \alpha$	$\sin \alpha \ / \ \cos \alpha$.
TGTSQD	$\tan^2 \theta$	
T2PHIB	$\tan 2\overline{\phi}$	

Flow Diagram

The flow diagram outlining the procedure that calculates the dip magnitude and azimuth and the degree and orientation of the resistivity anisotropy is on pages 332–334.

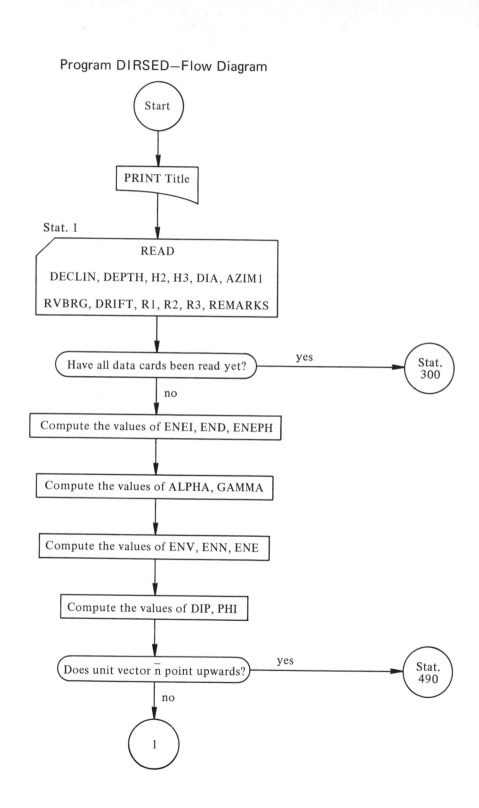

Program DIRSED—Flow Diagram

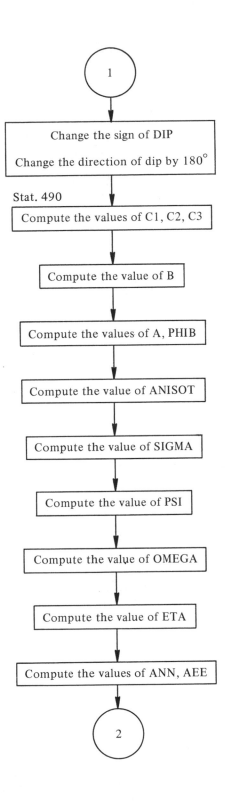

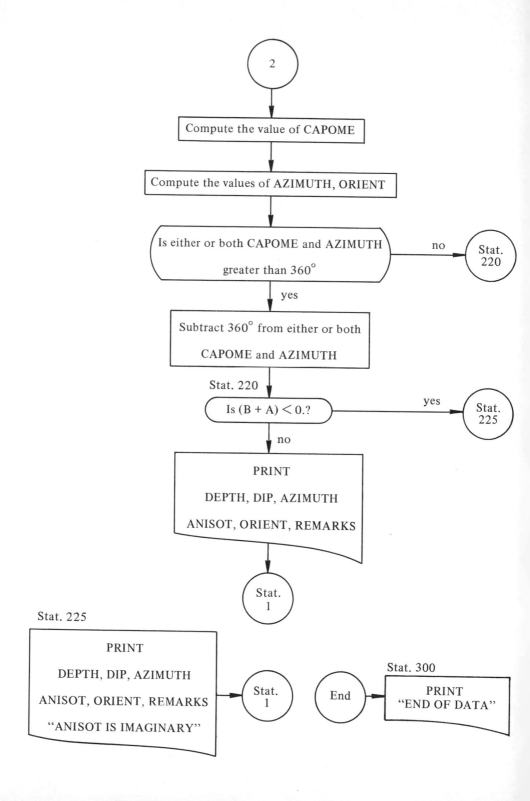

Computer Program

Using the nomenclature and flow diagram outlined in this section, a program in FORTRAN II language was prepared for making the required calculations with the CDC-6600 computer. The computer program is on pages 335–338.

PROGRAM DIRSED (INPUT, OUTPUT)

```
        PROGRAM DIRSED (INPUT, OUTPUT)
        PRINT 1000
1000    FORMAT (1H1)
        PRINT 1010
1010    FORMAT(7X,5HDEPTH,4X,3HDIP,3X,7HAZIMUTH,1X,7HANISOT.,1X,7HORIENT./
        17X,4HFEET,3X,7HDEGREES,1X,7HDEGREES,1X,7HA/B − 1,1X,7HDEGREES,10X,
        27HREMARKS/)
        READ 1015,DECLIN
1015    FORMAT(F10.2)
   1    READ 1030, DEPTH,H2,H3,DIA,AZIM1,RVERG,DRIFT,R1,R2,R3,REMARKS
1030    FORMAT(F10.2,9F7.2,A7)
        IF(DEPTH − 99999.) 2, 300, 300
   2    DP = H2+H3
        DM = H3−H2
        RADFAC = 3.1415927/180.
        TGTSQD = (DP**2+3.*DM**2)/2.25/DIA**2
        ENEI = 1./SQRTF(1.+TGTSQD)
        END = DP*ENEI/1.5/DIA
        ENEPH = DM*ENEI/0.866025/DIA
        DRIFT = DRIFT*RADFAC
        RVERG = RVBRG*RADFAC
        AZIM1 = AZIM1*RADFAC
        SB = SINF(RVBRG)
        CB = COSF(RVBRG)
        SD = SINF(DRIFT)
        CD = COSF(DRIFT)
        TGALPH = ABSF(SB/CD/CB)
        IF(SB) 14, 10, 22
  10    IF(CB) 11, 11, 12
  11    ALPHA = 3.1415927
        GO TO 30
  12    ALPHA = 0.
        GO TO 30
  14    IF(CB) 18, 20, 16
  16    ALPHA = 6.2831853−ATANF(TGALPH)
        GO TO 30
  18    ALPHA = 3.1415927+ATANF(TGALPH)
        GO TO 30
  20    ALPHA = 4.712389
        GO TO 30
  22    IF (CB) 26, 24, 28
  24    ALPHA = 1.5707963
        GO TO 30
  26    ALPHA = 3.1415927−ATANF(TGALPH)
        GO TO 30
```

```
 28    ALPHA = ATANF(TGALPH)
 30    GAMMA = AZIM1−ALPHA
       SG = SINF(GAMMA)
       GG = COSF(GAMMA)
       ENV = ENEI*CD+END*SD*CB−ENEPH*SD*SB
       ENN = −ENEI*CG*SD+END*(CG*CD*CB−SG*SB)+ENEPH*(−CG*CD*SB−SG*CB)
       ENE = −ENEI*SG*SD+END*(SG*CD*CB+CG*SB)+ENEPH*(−SG*CD*SB+CG*CB)
       DIP = ATANF(SQRTF(1.−ENV**2)/ENV)
       CPHI = ENN/SINF(DIP)
       SPHI = ENE/SINF(DIP)
 50    TGPHI = ABSF(SPHI/CPHI)
       IF(SPHI) 58, 52, 66
 52    IF(CPHI) 56, 55, 54
 54    PHI = 0.
       GO TO 80
 55    DIP = 0.
       PHI = 0.
       GO TO 80
 56    PHI = 3.1415927
       GO TO 80
 58    IF(CPHI) 62, 64, 60
 60    PHI = 6.2831853−ATANF(TGPHI)
       GO TO 80
 62    PHI = 3.1415927+ATANF(TGPHI)
       GO TO 80
 64    PHI = 4.712389
       GO TO 80
 66    IF(CPHI) 70, 68, 72
 68    PHI = 1.5707963
       GO TO 80
 70    PHI = 3.1415927−ATANF(TGPHI)
       GO TO 80
 72    PHI = ATANF(TGPHI)
 80    IF(ENV) 400, 490, 490
400    DIP = −DIP
       IF(PHI-3.1415927) 450, 480, 480
450    PHI = PHI+3.1415927
       GO TO 490
480    PHI = PHI−3.1415927
490    C1 = 1./R1**2
       C2 - 1./R2**2
       C3 = 1./R3**2
       B = 2.*(C1+C2+C3)/3.
       T2PHIB = ABSF(1.7320508*(C3−C2)/(2.*C1−C2−C3))
       PHIB = 0.5*ATANF(T2PHIB)
       IF(C3−C2) 108, 92, 98
 92    IF(2.*C1-B) 94, 94, 96
 94    PHIB = 0.
       A = 2.*C1-B
       GO TO 120
 96    PHIB = 1.5707963
       A = B-2.*C1
       GO TO 120
 98    IF(2.*C1-B) 104, 100, 102
100    PHIB = 2.35619
       A = 2.*(C2-C3)/1.7320508
       GO TO 120
```

```
102   PHIB = 1.5707963+PHIB
      GO TO 106
104   PHIB = 3.1415927-PHIB
106   A = 2.*(C3-C2)/1.7320508/SINF(2.*PHIB)
      GO TO 120
108   IF(2.*Cl-B) 116, 110, 112
110   PHIB = 0.785398
      A = 2.*(C3-C2)/1.7320508
      GO TO 120
112   PHIB - 1.5707963-PHIB
116   A = 2.*(C3-C2)/1.7320508/SINF(2.*PHIB)
120   ANISOT = SQRTF(ABSF((B A)/(B+A)))-1.
      IF(PHI-1.5707963) 130, 135, 135
130   SIGMA = PHI+4.712389
      GO TO 140
135   SIGMA = PHI-1.5707963
140   IF(AZIM1-SIGMA) 143, 150, 150
143   PSI = AZIM1-SIGMA+6.2831853
      GO TO 155
150   PSI - AZIM1-SIGMA
155   OMEGA = ATANF(ABSF(SINF(PSI)/COSF(PSI)/COSF(DIP)))
      IF(PSI-1.5707963) 180, 162, 164
162   OMEGA = 1.5707963
      GO TO 180
164   IF(PSI-3.1415927) 166,168, 170
166   OMEGA = 3.1415927-OMEGA
      GO TO 180
168   OMEGA = 3.1415927
      GO TO 180
170   IF(PSI−4.712389) 172, 184, 176
172   OMEGA = 3.1415927+OMEGA
      GO TO 180
174   OMEGA = 4.712389
      GO TO 180
176   OMEGA = 6.2831853−OMEGA
180   ETA = OMEGA+PHIB
      IF(ETA−6.2831853) 184, 182, 182
182   ETA = ETA−6.2831853
184   ANN = COSF(ETA)*COSF(SIGMA)-SINF(ETA)
      *COSF(DIP)*SINF(SIGMA)
      AEE = COSF(ETA)*SINF(SIGMA)+SINF(ETA)*
      COSF(DIP)*COSF(SIGMA)
      CAPOME = ATANF(ABSF(AEE/ANN))
      IF(ANN) 200, 186, 192
186   IF(AEE) 190, 188, 188
188   CAPOME = 1.5707963
      GO TO 210
190   CAPOME = 4.712389
      GO TO 210
192   IF(AEE) 198, 194, 210
194   CAPOME = 0.
      GO TO 210
198   CAPOME = 6.2831853-CAPOME
      GO TO 210
200   IF (AEE) 206, 202, 204
202   CAPOME = 3.1415927
      GO TO 210
```

```
  204   CAPOME = 3.1415927-CAPOME
        GO TO 210
  206   CAPOME = 3.1415927+CAPOME
  210   DIP = DIP/RADFAC
        AZIMUTH = PHI/RADFAC+DECLIN
        ORIENT = CAPOME/RADFAC+DECLIN
        IF(ORIENT-360.) 214, 212, 212
  212   ORIENT = ORIENT-360.
  214   IF(AZIMUTH-360.) 220, 215, 215
  215   AZIMUTH = AZIMUTH-360.
  220   IF(B+A) 225, 225, 230
  225   PRINT 1040, DEPTH, DIP, AZIMUTH, ANISOT,
        ORIENT, REMARKS
 1040   FORMAT(4X,5F8.2,3X,A7,2X,19HANISOT IS
        IMAGINARY)
        GO TO 1
  230   PRINT 1045, DEPTH, DIP, AZIMUTH, ANISOT,
        ORIENT, REMARKS
 1045   FORMAT(4X,5F8.2,3X,A7)
        GO TO 1
  300   PRINT 1050
 1050   FORMAT(5X,11HEND OF DATA)
        END
```

Appendix

Lithology Computer Program: Three Minerals
Plus Shale Contamination and Porosity
(Courtesy of M. M. Mosa)

Program Nomenclature

Symbol	Description
N	Number of involved equations.
Q	Control variable for the added mineral fractions.
P	Control variable for the first added mineral fraction.
AL	Upper limit value for the added mineral.
A	Matrix of the coefficient physical constants.
AA	Stored A.
kount	Control variable for title printings.
M	Control variable for depth levels.
LL	Control variable for plotting.
MM	Number of depth levels.
LQ	Maximum number of points to be plotted.
B	Variable parameters for log lookup.
BB	Stored B.
X	Computed porosity and mineral fractions.
XX	Stored X.
Y	Cumulative computed parameters.

Flow Chart

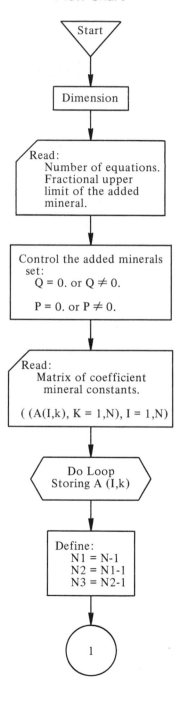

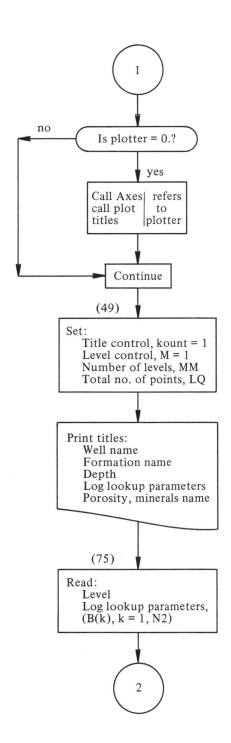

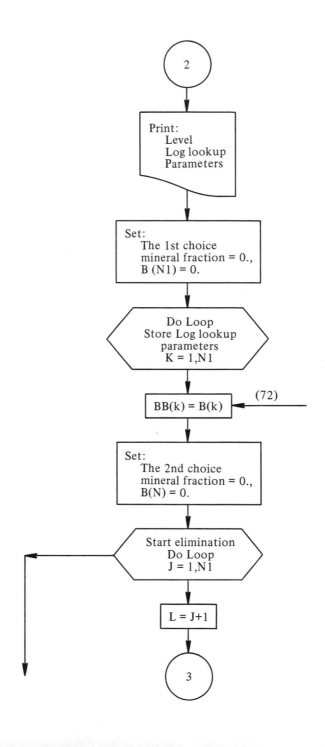

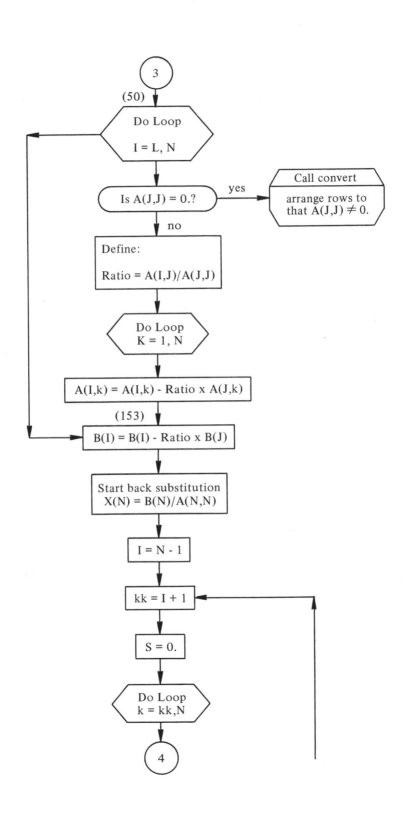

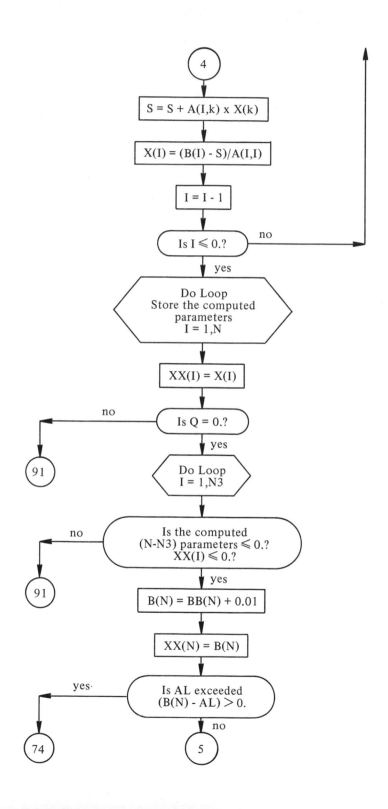

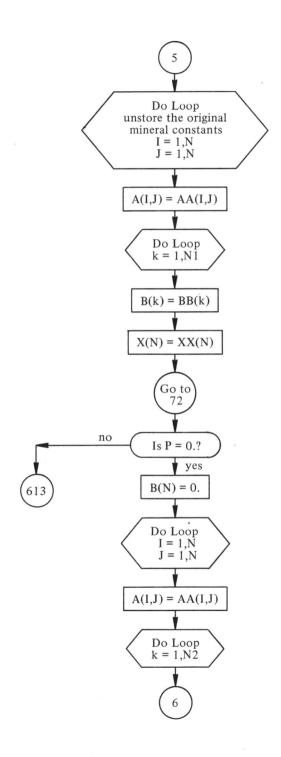

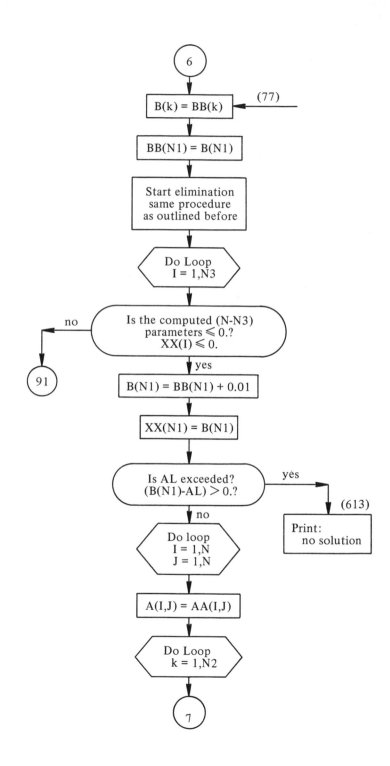

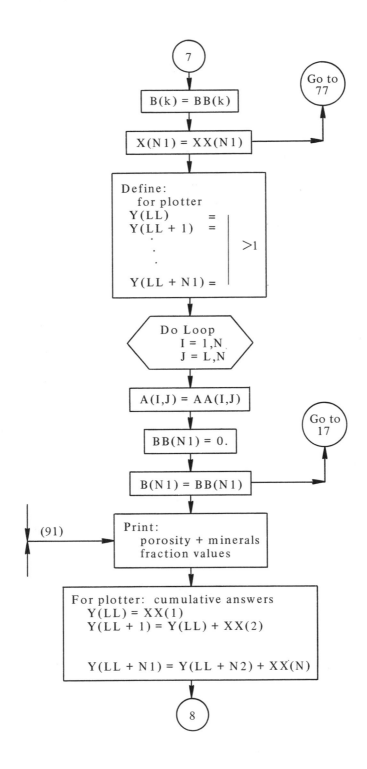

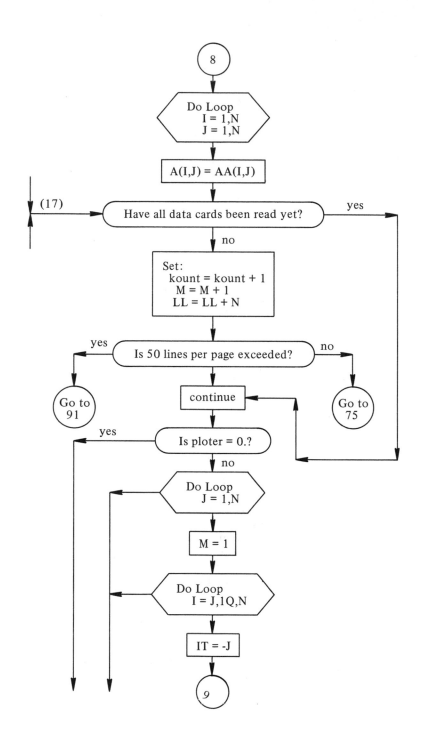

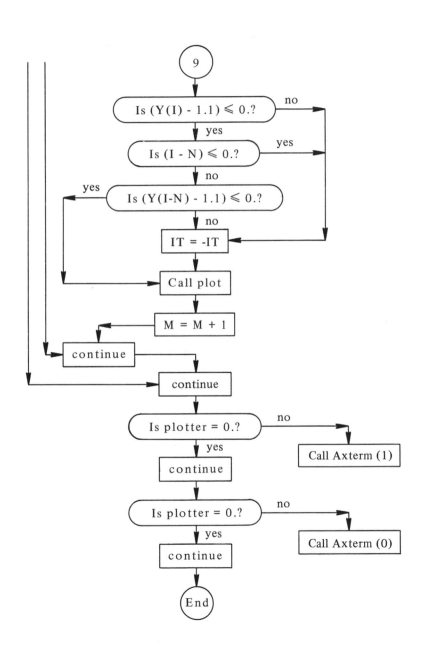

Subroutine Convert

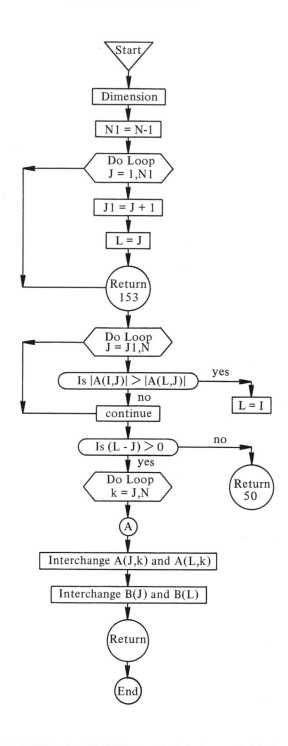

Program Listing

```
      PROGRAM PWELLOG (INPUT, OUTPUT)
      DIMENSION AA (10, 10),B(10),X(10),XX(10),A(10,10),BB(10),Y(4200),
    1 IC(10),, ICEL(10) , ICELL(10) , LEVEL1(700) , LEVEL2(700) ,
    1 ALEVEL(700)
      READ 5 , N , AL
    5 FORMAT (110 , F10.3)
      PRINT 12,N
   12 FORMAT (1H1,//6X,*N=*,13,//6X,* MATRIX DATA *//)
      Q = 0.0
    *  IF Q DOES NOT EQUAL TO 0.0 NO MINERAL FRACTION IS ADDED IN ORDER
    *  TO HAVE POSITIVE ANSWERS.
      P = 0.0
    *  IF P DOES NOT EQUAL TO 0.0 ONLY ONE ALTERNATIVE MINERAL FRACTION
    *  IS ADDED IN ORDER TO HAVE POSITIVE ANSWERS .
    *  IF PLOTTER = 0.0 , PLOT IS SUPPRESSED AND IF PLOTTER = 1.0 , PLOT
    *  IS MADE .
      PLOTTER = 1.0
      DO 100 I=1,N
      READ 150 , (A(I,K),K=1,N)
  100 PRINT 150 , (A(I,K) , K=1,N)
  150 FORMAT (6F10.3)
      DO 144 I=1,N
      DO 144 J=1,N
  144 AA(I,J) = A(I,J)
      N1 = N − 1
      N2= N1 − 1
      N3 = N2 − 1
      IF ( PLOTTER ) 200,201,200
  200 CALL AXES ( 42.0,7000.0,8200.0,1.0,8.0,0.0,1.0,60.,0.1,33,1500)
      IC(1) = 10H DAVIS E
      IC(2) = 10H WELL
      IC(3) = 10HNUMBER TWO
      ICEL(1) = 10H LOWER
      ICEL(2) = 10H CLEARFORK
      ICEL(3) = 10H FORMATION
      ICELL(1) = 10H       SAND
      ICELL(2) = 10H       GYPSUM
      ICELL(3) = 10H       SHALE
      ICELL(4) = 10H       ANHYDRITE
      ICELL(5) = 10H       DOLOMITE
      ICELL(6) = 10H       POROSITY
      IC1 = 8H0.00
      IC2 = 8H0.20
      IC3 = 8H0.40
      IC4 = 8H0.60
      IC5 = 8H0.80
      IC6 = 8H1.00
      IC7 = 8H7000
      IC8 = 8H7100
      IC9 = 8H7200
      IC10 = 8H7300
      IC11 = 8H7400
      IC12 = 8H7500
```

```
          IC13 = 8H7600
          IC14 = 8H7700
          IC15 = 8H7800
          IC16 = 8H7900
          IC17 = 8H8000
          IC18 = 8H8100
          IC19 = 8H8200
          CALL PLOTITL ( IC(1) , 10 , 2 , 3 , −3.5 , 5.5)
          CALL PLOTITL ( IC(2) , 10 , 2 , 3 , −3.5 , 5.0)
          CALL PLOTITL ( IC(3) , 10 , 2 , 3 , −3.5 , 4.5)
          CALL PLOTITL ( ICEL(1) , 10 , 2 , 3 , −3.5 , 3.5)
          CALL PLOTITL ( ICEL(2) , 10 , 2 , 3 , −3.5 , 3.0)
          CALL PLOTITL ( ICEL(3) , 10 , 2 , 3 , −3.5 , 2.5)
          CALL PLOTITL ( ICELL(1) , 10 , 2 , 2 , −1.3 , 7.32)
          CALL PLOTITL ( ICELL(2) , 10 , 2 , 2 , −1.3 , 5.99)
          CALL PLOTITL ( ICELL(3) , 10 , 2 , 2 , −1.3 , 4.68)
          CALL PLOTITL ( ICELL(4) , 10 , 2 , 2 , −1.3 , 3.33)
          CALL PLOTITL ( ICELL(5) , 10 , 2 , 2 , −1.3 , 2.00)
          CALL PLOTITL ( ICELL(6) , 10 , 2 , 2 , −1.3 , 0.67)
          CALL PLOTITL ( IC1 , 4 , 2 , 1 , −0.3 , 0.0 )
          CALL PLOTITL ( IC2 , 4 , 2 , 1 , −0.3 , 1.6 )
          CALL PLOTITL ( IC3 , 4 , 2 , 1 , −0.3 , 3.2 )
          CALL PLOTITL ( IC4 , 4 , 2 , 1 , −0.3 , 4.8 )
          CALL PLOTITL ( IC5 , 4 , 2 , 1 , −0.3 , 6.4 )
          CALL PLOTITL ( IC6 , 4 , 2 , 1 , −0.3 , 6.0 )
          CALL PLOTITL ( IC7 , 4 , 3 , 2 , 0.0 , −0.3 )
          CALL PLOTITL ( IC8 , 4 , 3 , 2 , 3.5 , −0.3 )
          CALL PLOTITL ( IC9 , 4 , 3 , 2 , 7.0 , −0.3)
          CALL PLOTITL ( IC10, 4 , 3 , 2 ,10.5 , −0.3 )
          CALL PLOTITL ( IC11, 4 , 3 , 2 ,14.0 , −0.3 )
          CALL PLOTITL ( IC12, 4 , 3 , 2 ,17.5 , −0.3 )
          CALL PLOTITL ( IC13, 4 , 3 , 2 ,21.0 , −0.3 )
          CALL PLOTITL ( IC14, 4 , 3 , 2 ,24.5 , −0.3 )
          CALL PLOTITL ( IC15, 4 , 3 , 2 ,28.0 , −0.3 )
          CALL PLOTITL ( IC16, 4 , 3 , 2 ,31.5 , −0.3)
          CALL PLOTITL ( IC17, 4 , 3 , 2 ,35.0 , −0.3 )
          CALL PLOTITL ( IC18, 4 , 3 , 2 ,38.5 , −0.3 )
          CALL PLOTITL ( IC19, 4 , 3 , 2 ,42.0 , −0.3 )
201       CONTINUE
          KOUNT = 1
          M = 1
          MM = 577
          LL = 1
          LQ = MM * N
49        KOUNT = 1
          PRINT 52
52        FORMAT ( 1H1 , /45X , 27HDAVIS E WELL NUMBER TWO)
          PRINT 33
33        FORMAT ( /45X,27HLOWER CLEARFORK FORMATION)
          PRINT 48
48        FORMAT (   //1X,116H   LEVEL   DEL-T   SUM   DEN
     1 SITY GRLOG POROSITY DOLOMITE ANHYDRITE SHALE GYPS
     1 UM SAND/)
75        READ 145 , LEVEL1(M) , LEVEL2(M) , (B(K) , K = 1,N2 ) , ISTOP
145       FORMAT ( 216 , F8.2 , F10.1 , F10.3 , F20.2 , 9X , I1 )
          PRINT 151 ,LEVEL1(M) , LEVEL2(M) , (B(K) , K = 1,N2 )
```

```
151   FORMAT ( 1H+ , 216 , 4F10.3 )
      ALEVEL(M) = ( LEVEL1(M) + LEVEL2(M) ) /2.0
      B(N1) = 0.0
      DO 146 K = 1 , N1
146   BB(K) = B(K)
      B(N) = 0.0
 72   BB(N) = B(N)
      DO 153 J=1,N1
      L = J+1
 50   DO 153 I = L,N
      IF (A(J,J) .EQ. 0.0) CALL CONVERT (A,B,N,J)
      RATIO = A(I,J) / A(J,J)
      DO 155 K=1,N
155   A(I,K) = A(I,K) = RATIO * A(J,K)
153   B(I) = B(I) − RATIO * B(J)
      X(N) = B(N) / A(N,N)
      I = N1
 70   KK = I + 1
      S = 0.0
      DO 63 K = KK,N
 63   S = S + A(I,K) * X(K)
      X(I) = (B(I) − S) / A(I,I)
      I = I −1
      IF ( I ) 83,83,70
 83   DO 89 I = 1,N
 89   XX(I) = X(I)
      IF (Q .EQ. 0.0) 61,91
 61   DO 80 I = 1,N3
      IF (XX(I) ) 950,950,80
 80   CONTINUE
      GO TO 91
950   B(N) = BB(N) + 0.010
      XX(N) = B(N)
      IF (B(N) − AL) 722,722,74
722   DO 723 I=1,N
      DO 723 J=1,N
723   A(I,J) = AA(I,J)
      DO 725 K=1,N1
725   B(K) = BB(K)
      X(N) = XX(N)
      GO TO 72
 74   IF (P .EQ. 0.0) 79,613
 79   B(N) = 0.0
      DO 414 I=1,N
      DO 414 J=1,N
414           A( I,J ) = AA( I,J)
      DO 413 K=1,N2
413   B(K) = BB(K)
 77   BB(N1) = B(N1)
      DO 154 J=1,N1
              L = J + 1
 51   DO 154 I=L,N
      IF (A(J,J) .EQ. O.O) CALL ADJUST (A,B,N,J)
      RATIO = A(I,J) / A(J,J)
      DO 156 K=1,N
156   A(I,K) = A(I,K) − RATIO * A(J,K)
```

```
  154   B(I) = B(I) − RATIO * B(J)
        X(N) = B(N) / A(N,N)
        I = N1
   71   KK = I + 1
        S = 0.0
        DO 64 K = KK, N
   64   S = S + A(I,K) * X(K)
        X(I) = (B(I) − S) / A (I,I)
        I = I − 1
        IF ( I ) 84,84,71
   84   DO 82 I = 1,N
   82   XX(I) = X(I)
        DO 81 I = 1,N3
        IF (XX(I) ) 951,951,81
   81   CONTINUE
        GO TO 91
  951   B(N1) = BB(N1) + 0.010
        XX(N1) = B(N1)
        IF (B(N1) − AL) 611,611,613
  611   DO 615 I=1,N
        DO 615 J=1,N
  615   A(I,J) = AA(I,J)
        DO 726 K=1,N2
  726   B(K) = BB(K)
        X(N1) = XX(N1)
        GO TO 77
  613   PRINT 302
  302   FORMAT (65X , 14H NO SOLUTION)
        Y(LL) = 1.2
        Y(LL+1) = 2.2
        Y(LL+2) = 3.2
        Y(LL+3) = 3.2
        Y(LL+4) = 4.2
        Y(LL+5) = 5.2
        DO 127 I=1,N
        DO 127 I=1,N
        DO 127 J=1,N
  127   A(I,J) = AA(I,J)
        BB(N1) = 0.0
        B(N1) = BB(N1)
        GO TO 17
   91   PRINT 161 , XX(1),XX(2),XX(3),XX(4),XX(5),XX(6)
  161   FORMAT (60X,F10.3,F10.3,F10.3,F10.3,F10.3,F10.3,
        Y(LL) = XX(1)
        Y(LL+1) = Y(LL) + XX(2)
        Y(LL+2) = Y(LL+1) + XX(3)
        Y(LL+3) = Y(LL+2) + XX(4)
        Y(LL+4) = Y(LL+3) + XX(5)
        Y(LL+5) = Y(LL+4) + XX(6)
        DO 128 I=1,N
        DO 128 J=1,N
  128   A(I,J) = AA(I,J)
        GO TO 17
   17   IF ( ISTOP .EQ. 0 ) 78,16
   78   KOUNT = KOUNT + 1
        M = M + 1
```

```
        LL = LL + N
        IF ( KOUNT . EQ. 51 ) 49,75
  16    CONTINUE
        IF ( PLOTTER ) 103,106,103
 103    Do 104 J = 1,N
        M = 1
        DO 104 I - J,LQ,N
        IT = - J
        IF ( Y(I) − 1.1 ) 113, 113,112
 113    IF ( I − N ) 112,112,110
 110    IF ( Y(I−N) − 1.1 ) 111,111,112
 112    IT = − IT
 111    CALL PLOT ( ALEVEL (M) , Y(I) , 1,IT )
        M = M + 1
 104    CONTINUE
 106    CONTINUE
        IF ( PLOTTER ) 105,107,105
 105    CALL AXTERM (1)
 107    CONTINUE
        IF ( PLOTTER ) 108,109,108
 108    CALL AXTERM(1)
 109    CONTINUE
        END
        SUBROUTINE CONVERT (A,B,N,J)
        DIMENSION A(10,10) ,B(10)
        N1= N − 1
        DO 153 J=1,N1
        J1 = J + 1
        L = J
 153    RETURN
        DO 40 I=J1,N
        IF (ABSF(A(I,J)) − ABSF(A(L,J))) 40,40,41
  41    L = I
  40    CONTINUE
        IF (L − J) 50,50,45
  50    RETURN
  45    DO 42 K = J,N
        TEST = A(J,K)
        A(J,K) = A(L,K)
  42    A(L,K) = TEST
        TEST = B(J)
        B(J) = B(L)
        B(L) = TEST
        RETURN
        END
        SUBROUTINE ADJUST (A,B,N,J)
        DIMENSION A(10,10) ,B(10)
        N1 = N − 1
        DO 154 J=1,N1
        J1 = J + 1
        L = J
 154    RETURN
        DO 40 I=J1,N
        IF (ABSF(A(I,J)) − ABSF(A(L,J))) 40,40,41
  41    L = I
  40    CONTINUE
```

```
        IF ( L – J ) 51,51,45
51      RETURN
45      DO 42 K = J,N
        TEST = A(J,K)
        A(J,K) = A(L,K)
42      A(L,K) = TEST
        TEST = B(J)
        B(J) = B(L)
        B(L) = TEST
        RETURN
        END
```

Appendix 3

Computer Program for Calculating
the Curvature of a Structure
(Courtesy of R. E. Muñoz-E)

```
      PROGRAM DERIV (INPUT,OUTPUT)
C     THIS PROGRAM EVALUATES THE SECOND VERTICAL DERIVATIVE OF
C     A STRUCTURAL SURFACE USING THE TEN RINGS AVERAGE METHOD.
C
C     THE IBM GRID VALUE DETERMINATION PROGRAM WAS USED TO DEFINE
C     THE REQUIRE GRID SYSTEM. THE GRIDS VALUES ARE FED
C     INTO THE COMPUTER AS INPUT DATA.
C
      DIMENSION TOP(83,126)
C
C     THE SECOND VERTICAL DERIVATIVE COEFFICIENTS ARE STORED
C     IN THE MEMORY.
      D0 = 2.82994
      D1 = -2.49489
      D2 = 0.05173
      D3 = -0.39446
      D4 = 0.00932
      D5 = -0.00732
      D6 = 0.00304
      D7 = 0.00219
C
C     THE DIVISORS WHICH CONVERT THE SUMS OF TOP TO AVERAGES
C     ARE INCORPORATED IN THE COEFFICIENTS.
      COEF0 = D0
      COEF1 = D1/4.
      COEF2 = D2/4.
      COEF3 = D3/8.
      COEF4 = D4/4.
      COEF5 = D5/8.
      COEF6 = D6/12.
      COEF7 = D7/12.
C
      PRINT 8
    8 FORMAT(5x,6HCOLUMN,4x,3HROW,7x,10HDERIVATIVE)
C
C     THE DATA ARE STORED IN THE MEMORY.
```

```
            DO 7 J = 1,126
            READ 1,MROW,NROW
    1       FORMAT (5x,14,1x,14)
    7       READ 2,(TOP(I,J),I = MROW,NROW)
    2       FORMAT (15F5.0)
C
C           THE SECOND VERTICAL DERIVATIVES ARE EVALUATED.
            DO 6 J = 9,117,4
            DO 6 I = 9,74,4
            ZERO = COEF0*TOP(I,J)
            ONE = COEF1*(TOP(I,J-1) + TOP(I,J+1) + TOP(I-1,J) + TOP(I+1,J))
            TWO = COEF2*(TOP(I-1,J-1) + TOP(I-1,J+1) + TOP(I+1,J-1)+
           1TOP(I+1,J+1))
            THREE = COEF3*(TOP(I-2,J-1) + TOP(I-2,J+1) + TOP(I-1,J+2) +
           1 TOP(I-1,J-2) + TOP(I+1,J+2) + TOP(I+2,J+1) + TOP(I+1,J-2) +
           2TOP(I+2,J-1))
            FOUR = COEF4*(TOP(I-2,J-2) + TOP(I-2,J+2) + TOP(I+2,J-2) +
           1TOP(I+2,J+2))
            FIVE = COEF5*(TOP(I+2,J-3) + TOP(I+3,J-2) + TOP(I+2,J+3) +
           1TOP(I-2,J+3) +TOP(I-3,J+2) + TOP(I-3,J-2) + TOP(I-2,J-3) +
           2TOP(I+3,J+2))
            SIX = COEF6*(TOP(I,J-5) + TOP(I+3,J-4) + TOP(I+4,J-3) + TOP(I+5,J)
           1+ TOP(I+4,J+3) + TOP(I+3,J+4) + TOP(I,J+5) + TOP(I-3,J+4) +
           2TOP(I-4,J+3) + TOP(I-5,J) + TOP(I-4,J-3) + TOP(I-3,J-4))
            SEVEN = COEF7*(TOP(I+1,J-7) + TOP(I+5,J-5) + TOP(I+7,J-1) + TOP(I+
           17,J+1) + TOP(I+5,J+5) + TOP(I+1,J+7) + TOP(I-1,J+7) + TOP(I-5,J+5
           2) + TOP(I-7,J+1) + TOP(I-7,J-1) + TOP(I-5,J-5) + TOP(I-1,J-7))
            DERIV = ZERO+ONE+TWO+THREE+FOUR+FIVE+SIX+SEVEN
    6       PRINT 10,J,I,DERIV
   10       FORMAT (5x,14,5x,14,5x,F10.1)
            END
```

Index

A

Abnormal pressure. *See* Pressure
Abnormally high pressure. *See* Pressure
Abyssal
  deposit, 184
  plain, 186
Accumulation, hydrocarbon. *See* Migration
Amplification factor, 290
Anisotropy
  ellipse, 119, 131-133
  fabric. *See* Fabric
  resistivity. *See* Resistivity
Anomaly
  geochemical, 208-211
  geothermal, 275-276
Aquifer
  common, 299, 302, 303, 306
  independent, 299
Archie formula, 305
Azimuth
  frequency diagrams, 141-145
  No. 1 electrode, of the, 85-90
  true, 97

B

Back-reef. *See* Reef
Barrier
  mechanical, 214
  permeability, 311, 316
  selective adsorption, 214
  sieving, 124
Basal schist, 180
Basin
  Delaware, 167, 170, 306
  Denver-Julesburg, 31, 32, 76-81, 306, 314, 322-327
  Maracaibo, 133
  Paris, 220
  Powder River, 306, 311
  San Juan, 306, 311
  Williston, 154, 157, 159, 220, 231-232
Bearing stress. *See* Stress
Bedding, graded, 76, 184
Bedrock, antochthonous, 218
Beds
  adsorption barrier, 212
  barrier, 212
  carrier, 212
  marker, 59-60
    X bentonite, 76-81
  Source
    areal extent, 33
    devolatilization, 30-33
    in mapping. *See* SP curve
    organic, 29
Bench
  sand, 63
  stairstepped, 67
Berm, 66
Big Horn Mountain, 311
Blairmore unconformity. *See* Unconformity

C

Capillary
  barrier, 318
  equilibrium, 296
  transition, 305
Catalyst, 5
Chalk
  Annona, 180, 195
  Austin, 180, 187, 189, 194-195,
    199, 201
  reservoir rock, 192
  Selma, 180, 195
Channel-fill sand bar. *See* Sand
Chemical potential, 286
Chenier ridges, 43
Chloride reacting value, 311
Chromatographic column, 215
  base exchange, 215
  heavy metallic ions, 214
  ion exchange, 215
  microgas analysis, 221, 233
  selective adsorption, 214
Clay minerals
  micas, 11
  stability realms, 12
Compaction, 208, 250, 251-253
  differential, 64, 77
  equilibrium, 255
  expelled water, 212
  model, 255-257
  normal, 250, 258
  of shales, 255-258. *See also* Shale
  Vertical, 257 *See also* Post com-
    paction; Precompaction
Computer applications, 161-173
  FORTRAN program, 329-338,
    339-356, 357-358
  dipmeter, 133
  flow chart, 340-350
  flow diagram, 331-334
  fracturing intensity, 201
  lithology, 169
  nomenclature, 329, 339
Concurrent deposition, 69
Consolidation
  model, 255-257
  theory, 255

Contact
  fresh salt water, 25-26
  gas water, 22, 23
  hydrocarbon water, 234. *See also*
    Hydrodynamics
  water-oil, 20-22, 292-303
Contour mapping, 110
Core holes, 95
Correlation
  auto, 182
  cross, 182
  curve, 182
  quality, 95
Coriolis force, 184
Cosmic rays, 218
Cross bedding, 92
Crossplot
  bulk density vs. neutron, 165
  sonic vs. bulk density, 163
  sonic vs. neutron, 164
Crude oil types, 28
Curvature, 197
  mapping, 181
  of structure, 198
Curves
  correlation, 90
  orientation, 90
  relative bearing, 90
  See also Eh; Lithology; SP curve 69
Cyclic subsidence, 69

D

Darcy's
  equation, 284-285
  law, 210
  type flow, 318
Deformation
  plastic, 130
  sensor. *See* Sensor
Delaware effect, 295
Delta
  environment, 49
  layering, 50-52
    bottom set, 50
    foreset, 50, 52
    prodelta, 50
    top set, 50
  margins, 29

Delta (cont.)
  sedimentation, 36-39
  sequence, 48-50
    constructional, 39
    destructional, 40
    directional, 50
Depositional
  fabric, 112
  petrofabric. *See* Petrofabric
  topography, 59, 61, 65
    surface, 80
    trough, 80
Depth of sealing, 273-274
Diagenesis
  epigenetic, 7
  locomorphic, 7
  of carbonate sediments, 8
  phylomorphic, 7
  redoxomorphic, 6
  sedimentary, 6
Diapiric shale, 255
Differential
  compaction. *See* Compaction
  travel-time, 262
Dilution patterns, 317-319
Dip
  angle, 85
  patterns, 105
  regional, 102
  true, 97
  vectors, 140
Dipmeter
  CDM-P, 88
  CDM-T, 87
  continuous, 119-152
  FORTRAN program. *See* Com-
    puter applications
  four arm, 88, 94
  high density results, 140-149
  poteclinometer, 88-89
  resistivity, 86
  Seiscor. *See* Seiscor dipmeter log
  SP, 86
  teleclinometer, 87-88
  three arm, 119
  wire line, 86
Directional tectonic trends. *See* Tectonic
  resistivity. *See* Resistivity

Dolomite
  Ellenburger, 220
  Fusselman, 220
  Smackover, 187
Donnan equilibrium, 321
Drainage pattern, 61
Draping, 69
Drift, 86, 90
Drilling muds
  fresh water, 21
  organic, 22
  salt, 5
  sea water, 21
Driving potential, 285

E

Earthquake
  Grand Banks, 183. *See also* Turbi-
    dite
  frequency, 188
Eccentricity, 125
Eh
  curve, 1-2
  intensity level, 5
  poising capacity, 5
  zero line, 4
Eh-pH diagrams, 7, 9
  fence diagram, 10
  of natural environment, 9
Electric potential, 288
Electrode
  calomel, 4
  gold, 22
  hydrogen, 4, 33
  inert, 5
  iron, 22
  noble metal, 22
  orientation, 85
  reference, 22
    moving, 26
    unstable, 26
Electro-osmosis, 306
en echelon, 67
Environment
  continental shelves, 12
  geochemical, 2
  marine sedimentation, 9

Equations
  bulk volume, 160
  Darcy's. *See* Darcy's
  density, 160
  hydrogen index, 160
  identity, 161
  Kozeny's, 125
  Laplace, 199
  neutron, 160
  saturation, 160
  simultaneous, 160-161
  sonic, 160
  Van't Hoff, 320
Equatorial plane, 98
Equipotential surface, 289
Ethane gas. *See* Soil air
Evaporite sediments
  insoluble salts, 171
  potash minerals, 171
  sodium salts, 171
Exshaw unconformity. *See* Unconformity

F

Fabric
  acoustic, 123
  anisotropy, 112
  apposition, 119
  method
    aggregate, 122-123, 125-127
    dielectric, 123
    particulate, 120-122
    resistivity, 125-127
  primary, 119
  sand, 120-123. *See also* Sand.
  secondary, tectonic. *See* Tectonic
  sedimentary rocks, 119
Facies, 60
  abyssal, 64, 74
  delta-platform, 73
  paleo-. *See* Paleo-
  sedimentary. *See* Sedimentary
    facies
  shelf, 64
Failure
  brittle, 188
  by flexing, 188

Fault
  cantilever, 188-189
  contemporaneous, 105
  depositional, 105
  detection, 187-188
  displacement, 180
  downthrown, 195
  drag, 105, 189
  gravity (growth), 105, 113, 188
  normal tension, 105
  overthrust, 113
  recognition, 181-184
  roll-over, 105, 188-189
  sealing, 299
  shattering, 180
  strike-slip, 139
  tension, 180
  throw, 105
  upthrown, 195
  up-to-the-coast, 189
  wrench, 131-139
  zone, 189
    Luling-Mexia, 187-195
    Mexia-Talco, 159
    Urdaneta, 133-139
Field
  Adena, 314
  Alida, 159
  Aneth, 220
  Antelope, 180
  Bell Creek, 70
  Black Jack, 317-319, 326-327
  Branyon-Buchanan, 193, 203
  Chanute, 72
  Chaveroo, 159
  Coulommes, 220, 222-223
  Coyote Creek, 70, 313-314
  Donkey Creek, 313
  Florence, 180
  Glenburn, 220, 231
  Guselky, 211
  Halverson, 313
  Hassi-Messaoud, 74
  Hilbig, 212, 233-239
  Innisfail, 220, 224-225
  Kummerfeld, 313
  Lake Maracaibo, 70
  Leader, 315

Field (cont.)
Levelland-Slaughter, 159
Luling-Branyon, 203
Miller Creek, 313
Parkman, 159
Pearsall, 199
Pine Island, 180
Pratt, 220, 231
Salt Creek, 313, 314
Salt Flat, 180, 193-194, 203
Seligson, 74
Steelman, 159
Tenney Creek, 193-194
Wellman, 221, 230
Westerose, 220, 228-229
Weyburn, 159
Wiley, 159
Wizard Lake, 220
Flow potential, 285-286, 289
amplified-gas, 290
amplified-oil, 290
amplified-water, 290
gradient, 284
minimum, 289
Fluid
migration. *See* Migration
motion, law of, 284
transfer, 285
transport, 285
Flushing
fresh water, 316-317
hydrodynamic, 306-319
Force
Coriolis. *See* Coriolis force
fields
gravity, 120-121
magnetic, 120-121
Fore-reef. *See* Reef
Formation
pressure, estimation of, 258-279
method
acoustic travel-time, 258-264
shale density, 272-274
shale resistivity, 264-271
waters
associated crudes, 26
carbonate and uranium, 27
infiltration, 26

oil field, 27
polluted, 26
surface, 27
Formations
Atoka, 48
Bathonian, 220, 222-223
Callovian, 220, 223
Cliff House, 42
Dewey Lake, 221
Ellenburger, 173-175
Ireton, 173-175
Leduc reef, 224-229
Lewis, 42
Madison, 157-159
Mancos, 42
Mesaverde, 42
Mission Canyon, 157
Point Lookout, 42
San Andres, 159, 175-177
Smackover, 159
Spearfish, 157, 220, 231-232
FORTRAN programs. *See* Computer
applications
Fracture
finding, 190
index
density, 197-202
intensity, 189-196
proximity, 196-197
intensity, 112, 180, 188-189
natural, 180
occurrence, 197
maximum probability, 197
trend analysis, 199
porosity, 190-191
Fracturing
gradient, 278
intensity FORTRAN program. *See*
Computer applications
Fresh salt water contact. *See* Contact

G

Gas water contact. *See* Contact
Geostatic
equilibrium, 213, 250-283
pressure, 256-257, 275
Graded bedding. *See* Bedding

Grain orientation, 112, 117
Granulometry, 219

H

High density computation, 101
Hydraulic leakage, 208-211, 215, 217
Hydrocarbon Migration. *See* Migration
Hydrocarbons, detection of, 33
Hydrodynamic, 286
  desalting, 320
  entrapment, 208. *See also* Mapping
    mapping of, 291-294
  flushing. *See* Flushing
  trap delineation, 288-291
Hydrodynamics, 284-328
  of compaction, 291
  of infiltration, 284
  measuring devices, 288
  potential, 288
  water-hydrocarbon contact, 284-288
Hydrogen sulfide, 26, 33
Hydrogeology, 207
  paleohydrodynamics, 208
  periods, 208
Hydro-osmosis, 285
Hydro-osmotic oil entrapment. *See* Oil
Hydrostatic
  equilibrium, 318
  pressure, 250, 251
    fluid, 257
    gradient, 251
    normal, 251

I

Index
  density. *See* Fracture
  intensity. *See* Fracture
  proximity. *See* Fracture; Reef
    warping. *See* warping
    See also Regression,
Induced porosity. *See* Porosity

Infiltration, 284
  laterally, 284
  vertically downward, 284
Ion sieve, 277
Isochronal distance, 60
Isopach, 56
  map, 61
  selective, 64, 80
Isosalinity, 314

K

Kozeny's equation. *See* Equation

L

Laccolith, serpentine, 221
Laplace equation. *See* Equation
Lateral
  compression, 130
  tectonic stresses, 130
Limestone
  Arbuckle, 286
  Buda, 189
  Deep Edwards, 299, 301
  Edwards, 189, 199
  Mission Canyon, 220, 232
Lithofacies, 153
Lithology
  FORTRAN program. *See* Computer applications
  logging, 159-173
    two porosity curves and gamma ray, 169-170
    three porosity curves, 162-169
Loading depression, 70
Log
  open-circuit SP, 19-25
  redoxomorphic. *See* Redoxomorphic
  three porosity. *See* Three porosity logs
Longshore currents, 143

M

Magnetic
  declination, 85
  north, 85

Mapping
  contour. *See* Contour mapping
  hydrodynamic entrapment, 291-295
    amplified water flow potential, 294
    minimum oil flow potential, 294
    sequence of, 291, 292
    structure, 111, 291, 292
    water flow potential, 292, 293
  index. *See* Fracture
  parameters, 207
  radioactivity. *See* Radioactivity mapping
  source beds. *See* SP curve
Maracaibo Lake, 134
Marker
  beds. *See* Beds
  paleo-electric, 187
  Textularia hockleyensis, 187
Membrane
  ion exchange, 319
  permoselective, 319
  semipermeable, 321
Merging effects, 217
Meridian line, 98-99
Methane gas. *See* Soil air
Migration
  fluid, 213
  hydrocarbon, 207
  primary, 211, 217
  release mechanism, 211
  secondary, 211, 218
  waterborne, 207
  water expulsion, 207
Mincral
  facies, 162
    eight-mineral, 167
    three-mineral, 162
    two mineral, 162
  log responses, 156
Mineralization effects, 210
  anomalies, 212
  chimney, 214, 217, 221, 229
  disturbing factors, 217
  fossil chimney, 219
  funnel, 217, 218, 221, 229
  inorganic, 213
  organic, 212, 213, 214

Minerals. *See* Clay.
Mud flow, 182

N

Nomograph, 166
Normal density–compaction gradient, 274

O

Offshore sand bar. *See* Sand
Oil
  accumulation, downward, 51
  entrapment, hydro-osmotic, 326
  -water contact. *See* Contact
  zone, 20
Olistostrome, 184
Open battery potential, 16
Open-circuit SP log. *See* Log
Osmo-dynamic potential, 287
Osmosis, 285-288
  chemi, 286-288
  electro, 286-288
  hydro, 286-288
  thermo, 286-288
Osmostatic equilibrium, 322
Osmotic
  flow, 318-326
    fluid transfer, 326
    pressure, 325
  force, 287
  pressure barrier, 277, 279, 306, 323-327
Overburden pressure. *See* Pressure
Ovcrcompaction, 277
Overpressure, 250, 251
Oxidation levels, 16

P

Packing, 120
Paleo-facies, 153, 173-175
  distribution. *See* Sand
  strandline, 54
Paleodepositional
  strike, 68
  surface, 67
  trend, 66

Paleostream, 61
Particle
  disk, 120
  rod, 120
  shape, 120
Pelagic sediments, 186
Permeability
  barrier. *See* Barrier
  directional, 119
  ratio, 125
Petrofabric, 37
  orientation, 119, 127-129
  primary, 112
  secondary, 112, 116
  tectonic induced. *See* Tectonic
Petrophysical
  properties, 123
  tensor, 123-124
Photoclinometer, 86-87
pH or (proton) activity, 2, 3
Physicochemical properties
  oil wet, 216
  oxidation, 208
  redox potential. *See* Redox
  reduction, 208
Pinch-out, up-dip, 153
Plastic flow, 112
Plow pad, 94
Poisson ratio, 279
Polar
  coordinates, 131
  diagrams, 141
Pore pressure. *See* Pressure
Porosity
  block, 191
  by deformational shattering, 189
  curves. *See* Lithology logging
  fracture, 189-196, 198-199
  induced, 112, 191
  matrix, 191
  of clay, 256
  of shale, 257-258, 271, 272
  partitioning coefficient, 189
  vuggy, 189
  See also Three porosity logs
Poteclinometer, 88-89
Potentiometer, slide wire, 90
Post compaction, 69

Potash deposits, 171-172
Potassium, abnormal concentration
  in, 217
Precompaction, 69
Pressure, 208
  abnormal, 274-279
  abnormally high, 250, 264
  geostatic, 209
  hydrostatic, 209
  osmotic. *See* Osmotic
  overburden, 209, 256
  pore, 319. *See also* Shale
  shale. *See* Shale
Pressure seal, 275
Propane gas. *See* Soil air
Pyrolization, 211

Q

Quartzose rocks, 311

R

Radioactivity mapping, 207-239
  background level, 215
  correction charts,
    for cased hole, 243
    for hole size, 242
    for thin bed, 244
  elements, 216
  fallout, 218
  normalization, 217-218
    thick bed, 241
    thin bed, 242
  soil samples, 234
  techniques, 241-245
  unconformities, 217
  unit conversion factors, 245
Radium, 216
Reactions, Eh and pH dependent, 3
Redox (reduction-oxidation)
  gradient, vertical, 30
  potential, 1-2, 51, 233, 238
    physicochemical properties, 207,
      233
  probe, 233
    electrode, 233
    of soil sample, 240
    surveys, 234

Redox (cont.)
    well log, 17-19, 22
        in mapping petroliferous environ-
            ment, 29-30
        in mapping source beds, 29
        in oil and gas exploration, 28-30
Redoxomorphic
    log, 1
    sequence, 60
    series, 12-13
Reef
    back, 29, 153
    bank, 154
    build-up, 153
    core, 29
    fore-, 29, 153-154
    hinge line, 154
    patch, 154
    proper, 153
    proximity index, 175
Regional dip. See Dip
Regression, 36-37, 38
    accelerated, 39, 41
    complex patterns, 45
    decelerated, 39, 41
    index, 56
    linear, 39, 41
    patterns, 37-42
    shore line, 54
    stable, 53
    strandline trap, 51
Relative bearing. See Curves
Reservoir
    closure, 299, 301
    zonation. See Zones
Resistivity,
    anisotropy, 119, 127, 131. See also
        Fabric
    directional, 125-127
    Ro, 125
    Rw, 125
    shale. See Shale
Rwa function, 306
Rwe function, 318-319, 322, 326

                    S

Salinity
    patterns

breakthrough, 314
    dilution, 313
    shadow effect, 313
Salt
    dome, 130
        deep-seated, 130
        Moss Hill, 113
        piercement, 112, 130
    overhang, 110
Sand
    bar-finger, 74-75
    barrier island, 66
    beach, 70
    bench, 63
    body development, 65
    channel, 69
    chenier, 71
    delta, 72-75
    distribution patterns, 59
    fabric orientation, 126-127
        alluvial, 128
        barrier bar, 141
        beach, 128
        channel-fill, 36-37, 38, 46-47,
            142
        continental, 128
        deep marine, 128
        offshore, 128, 141
        tidal channel, 142
        trough filled, 142
    levies, 74
    lunate bar, 74
    offshore bar, 36, 37, 38, 43-46, 66
    paleo-distribution, 60
    pinch-out, 64
    point-bar flood plain, 70
    proximity index, 175
    shoreline, 70
    strike valley, 65
    turbidite, 29, 74-76
    zones. See Zones
Sands
    Booch, 74
    Channel fill, 68-69
    Cherokee, 66
    Chester, 74
    Cockfield, 113, 115, 185
    Dakota, 313-315

Sands (cont.)
 D and J, 31-33, 317-319, 322-327
 Delaware, 48, 76
 First Wall Creek, 311-312
 Hoover, 43-44
 Jackson, 40
 Morrow, 71, 195
 Olmos, 199-200
 Oriskany, 43
 Rotliegend, 234, 240
 Sanish, 180
 Santa Barbara, 133
 Shoestring, 67
 Spraberry, 180, 195
 St. Peter, 43
 Tonkawa, 43-44
 Topanga, 184
 Tuscaloosa, 22, 23
 Viking, 41
 Wilcox, 299-300
 Yegua, 185
Schmidt diagram, 144-149
Scour Channel, 67
Sealed morphological entity, 64
Secondary petrofabric. *See* Petro-
 fabric
Second derivative
 space, 197
 vertical, 199-202
Sedimentary facies,
 anhydrite sheet, 154-155
 biogenic, 153-154
 edge
  porosity, 158
  sealing, 158
 genetic process, 154
 lateral, 154
 marker bed, 158
 time regressive, 158
Sedimentation
 cycles, 36
 delta. *See* Delta
 environment, 13-17
  littoral, 12
  neritic, 12, 13
  sublittoral, 12
 patterns, 37-50
  markers, 182

 message, 182
 random noise, 181
 recurring, 181
 sequence,
  shoreline
   regressive, 13-14
   transgressive, 15-17
Seiscor dipmeter log
 computer, 100-102
 reader, 96-97
Selective solubilization, 211
Semipermeable membrane, 277, 286,
 318-323
Sensor
 deformation, 195
 remote, 214-215
Serpentine plug, 113
Shale
 baseline shift, 16
 compaction model, 256
 correlatable sequence, 181
 density, 272-274
  by calibrated fluid column, 272
  cutting, 272
  dehydration, 278
  See also Formation
 masses, deep-seated, 131
 pressure, 250
  from shale density, 250
  from shale resistivity, 250
  from shale sonic transit time, 250
  pore, 254, 274, 275
 resistivity, 253-255
  compaction trend, 265
  hypothetical, 271
  normal trend, 254
  See also Formation
 water, 254
  composition, 253-254
  density, 316
  encroachment, 278
  influx, 278
  resistivity, 316
Shales
 Callovian, 220, 223
 Calmar, 229
 Dakota, 180
 Exshaw, 221, 225, 229

Shales (cont.)
  Ireton, 227, 229
  Jackson, 185
  Joli Fou, 229
  Lea Park, 229
  Paradox, 220
  Taylor, 186-187
  Woodford, 219
Shalyness factor, 322
Shelf margins
  carbonate, 29
  sandy, 29
Shoreline
  characteristics
    regressive, 37
    stable, 37
    transgressive, 37
  sand. *See* Sand
Slump blocks, 184-187
Soil air
  ethane gas in, 236
  methane gas in, 235
  propane gas in, 237
Solution
  external, 321
  internal, 320
Source beds. *See* Beds
Specific surface, 126
SP curve, 22, 288
  fingers, 37-50, 181
  kicks, 181
  in mapping source beds, 30-33
  plot, 316
  shapes, 37-50
Standstill, 70
Static equilibrium, 291
Stereographic net, 98-100
Strain
  elliptic, 112
  rotational, 112
  vectors, 113, 117
Strandline
  paleo-. *See* Paleo-
  transgression, 57
  trap. *See* Regression
Strata
  genetic interval of, 61, 80
  genetic sequence of, 60

Stratigraphic
  markers, 61
  traps, 50-81 128, 209, 318
Stress
  bearing, 256
  compressive, 257
  effective, 256
Structural
  dip, 102
  elevation, 290
  position, 64
Structure mapping. *See* Mapping
Sulfur
  content log, 172
  Delaware Basin, 170
  Native, 170
  Sicilian, 170
Suspended material, 183
  cinder clouds, 183
  snow avalanche, 183
  turbidite, 183
  turbulence, 183
Symmetry
  diverging, 45, 46
  horizontal, 37
  parallel, 37

T

Tectonic
  deformation, 129
  induced petrofabric, 114, 116
  secondary fabric, 119, 129
  trends, directional, 112
Telluric current, 234, 286
Terzaghi—Peck model, 255-257
Thermal potential, 289
Thickness, true, 104
Three porosity logs, 159-173
Thrust faulting, 277
Tidal
  current, 127
  pulsations, 188
Tide, terrestrial, 188
Time
  line, 48
  markers, 59-61

Transgression, 36-38
  accelerated, 43
  complex patterns, 45
  decelerated, 43
  linear, 43
  patterns, 42-43
  strandline. *See* Strandline
Turbidite, 36-38, 47-48, 74-76,
  127-128, 184-187
  Grand Banks, 75. *See also* Earth-
    quake
  sand. *See* Sand
Turbodrilled holes, 95

U

Unconformity, 110, 113, 131
  Blairmore, 221
  Exshaw, 221
Undercompaction, 258, 277
Up-dip, pinch-out. *See* Pinch-out
Uranium, halflife of, 218
Urano-organic compounds, 216

V

Van't Hoff equation. *See* Equation
Vertical redox gradient. *See* Redox

W

Warping, 112, 113
  index (degree of), 129
Water
  escape, vertical, 275

  expulsion. *See* Migration
  invasion, 311
  motion
    down-dip, 319
    up-dip, 318
  -oil contact. *See* Contact
  saturation
    downward projection, 296-303
    normalized, 301, 306
  zone, 20
Waterborne process, 211
  primary, 216, 217
  See also Migration
Wave
  amplitude
    attenuation, 190
    reduction, 190
  compressional, 190
  energy, 37
  shear, 190
  sonic, 190

Z

Zenith point, 98
Zero reference, 33
ZoBell
  communication, 19
  solution, 4
Zones
  fault. *See* Fault
  reservoir, 296-299
    independent, 307-310
  sand, 303
  water. *See* Water